Praise for

Essential Hempcrete Construction

...Hemp as a building material is one of the most important uses of this plant. *Essential Hempcrete Construction* explains this new technology in a clear and concise manner. With this information anyone wanting to use this material will have a great starting point.

—STEVE ALLIN, Director, International Hemp Building Association and author, *Building with Hemp*

In a field that is often beset with more wishful thinking than honest appraisal, *Essential Hempcrete Construction* stands out as a well-researched in investigation into the potential and promise of hemp-lime systems in the contemporary built-environment.

—TIM CALLAHAN, co-author, *Building Green* and founding partner at Alembic Studio, LLC

Hempcrete/hemp-lime construction is one of the simplest and most effective and affordable solutions for sustainable, healthy and energy efficient building available to mainstream builders and self-help enthusiasts alike. Magwood has years of experience of natural building and sound construction practice so this book provides excellent, well-illustrated, guidance.

—TOM WOOLLEY, author, *Building Materials, Health and Indoor Air Quality; Low Impact Building;* and *Hemp-Lime Construction*

Essential Hempcrete Construction is an informative and easy to read guide/instruction manual to building with hemp-lime. Magwood has been able to simplify the data for us and compare costs to other conventional building materials, he has also managed to keep it brief and to-the-point so as to quickly help us make an informed decision to build with hemp-lime or not. This book is a time (and money) saver - worth every penny.

—GREG FLAVALL, Hemp Technologies Collective

Essential Hempcrete Construction is an exciting manual for homeowners, building teachers, building officials and contractors. Filled with photos and drawings that help all of us make choices that are ultimately going to support long lasting buildings. Solid and fun to read.

—SUKITA REAY CRIMMEL, coauthor, *Earthen Floors*

Chris Magwood has combined his deep understanding of building science with some fortuitous hands-on experience of working with hempcrete to craft this timely and detailed guide to the essentials of hempcrete construction. By simply mixing the lightweight core of hemp stalks (an agricultural byproduct) with lime it is possible to make an insulating material that can handle moisture without decomposing, has good structural qualities and thermal performance, is nontoxic and fire resistant, naturally sequesters carbon, and is ultimately completely recyclable. With the successful use of hempcrete in Europe for over a decade, hopefully this book will help usher in a new era of industrial hemp production in North America.

—KELLY HART, www.greenhomebuilding.com

Chris Magwood's easy-to-read, thoughtful, and practical book on building with hempcrete is a must-have for my work as a designer and builder focused on resilient, natural, and low carbon construction. In a world increasingly threatened by climate change, practical solutions are needed now to lower our collective ecological impact as a building trade. *Essential Hempcrete Construction* is a critical tool in our toolkits.

—ACE MCARLETON, coauthor, *The Natural Building Companion*, and cofounder, New Frameworks Natural Design/Build LLC

It's amazing to me that, in the 21st century, we still have to educate people about the many uses for industrial hemp. With this book, Chris Magwood presents an irrefutable endorsement for hemp. It's simple common sense for building a better future—literally!

—DAN SKYE, editor-in-chief, *High Time*s magazine.

Essential Hempcrete Construction provides an excellent pathway towards creating buildings that are more ecologically sound, healthful, and beautiful. This guide is amazingly comprehensive – materials, techniques, costs, building science, illustrations, recipes, pros and cons and more – yet concise and readable. Great information for experienced natural builders as well as do-it-yourself neophytes.

—GAYLE BORST, Registered Architect, Pres/CEO of Stewardship, Inc., and founding member, Design~Build~Live

essential HEMPCRETE CONSTRUCTION

the complete **step-by-step** guide

Chris Magwood

Cover design by Diane McIntosh.
All photos © Chris Magwood, unless otherwise noted. Illustrations © Dale Brownson
Thumbs up Art: AdobeStock_23490949

Printed in Canada. Third printing September 2022.

This book is intended to be educational and informative. It is not intended to serve as a guide. The author and publisher disclaim all responsibility for any liability, loss or risk that may be associated with the application of any of the contents of this book.

Funded by the Government of Canada | Financé par le gouvernement du Canada | Canada

Paperback ISBN: 978-0-86571-819-7
eISBN: 978-1-55092-613-2
Series ISBN: 978-0-86571-821-0

Inquiries regarding requests to reprint all or part of *Essential Hempcrete Construction* should be addressed to New Society Publishers at the address below. To order directly from the publishers, please call toll-free (North America) 1-800-567-6772, or order online at www.newsociety.com

Any other inquiries can be directed by mail to:
New Society Publishers
P.O. Box 189, Gabriola Island, BC V0R 1X0, Canada
(250) 247-9737

New Society Publishers' mission is to publish books that contribute in fundamental ways to building an ecologically sustainable and just society, and to do so with the least possible impact on the environment, in a manner that models this vision. We are committed to doing this not just through education, but through action. The interior pages of our bound books are printed on Forest Stewardship Council®-registered acid-free paper that is **100% post-consumer recycled** (100% old growth forest-free), processed chlorine-free, and printed with vegetable-based, low-VOC inks, with covers produced using FSC®-registered stock. New Society also works to reduce its carbon footprint, and purchases carbon offsets based on an annual audit to ensure a carbon neutral footprint. For further information, or to browse our full list of books and purchase securely, visit our website at: www.newsociety.com

Library and Archives Canada Cataloguing in Publication

Magwood, Chris, author
Essential hempcrete contruction : the complete step-by-step guide / Chris Magwood.

(Sustainable building essentials)
Includes bibliographical references and index.
Issued in print and electronic formats.
ISBN 978-0-86571-819-7 (paperback).--ISBN 978-1-55092-613-2 (ebook)

1. Hemp. 2. Concrete. 3. Plant fibers as building materials.
4. Aggregates (Building materials). I. Title.

TA455.P49M33 2016 693'.997 C2016-902871-2
C2016-902872-0

Contents

New Society Sustainable Building Essentials Series

Series editors
Chris Magwood and Jen Feigin

Title list

Essential Hempcrete Construction, Chris Magwood

Essential Prefab Straw Bale Construction, Chris Magwood

Essential Building Science, Jacob Deva Racusin

See www.newsociety.com/SBES for a complete list of new and forthcoming series titles.

THE SUSTAINABLE BUILDING ESSENTIALS SERIES covers the full range of natural and green building techniques with a focus on sustainable materials and methods and code compliance. Firmly rooted in sound building science and drawing on decades of experience, these large-format, highly-illustrated manuals deliver comprehensive, practical guidance from leading experts using a well-organized step-by-step approach. Whether your interest is foundations, walls, insulation, mechanical systems or final finishes, these unique books present the essential information on each topic including:

- Material specifications, testing and building code references
- Plan drawings for all common applications
- Tool lists and complete installation instructions
- Finishing, maintenance and renovation techniques
- Budgeting and labor estimates
- Additional resources

Written by the world's leading sustainable builders, designers and engineers, these succinct, user-friendly handbooks are indispensable tools for any project where accurate and reliable information are key to success. GET THE ESSENTIALS!

Acknowledgments

I AM EXTREMELY LUCKY AND GRATEFUL that the "rock stars" of the natural building world are simultaneously true superstars of forward thinking about the care of people and the planet *and* humble, approachable, brilliant and most often hilarious people who have built an extended family that I value greatly.

I am equally blessed to have a supportive and caring and patient and engaged immediate family. Together, they have made me who I am, and there aren't really thanks enough for that.

This talk is like stamping new coins.
They pile up,
while the real work is done outside
by someone digging in the ground.

— *Rumi*

Chapter 1

Introduction

IN 1998, a ban on the cultivation of industrial hemp in Canada was ended with the passing of the *Industrial Hemp Regulations.* Shortly thereafter, I received a call from a neighboring farmer, Grant Moorcroft. He was aware that I had been building with straw bales for a few years, and he wanted me to come and check out the bales of hemp straw he had made from his first hemp crop and see if they would be useful as building bales.

As it turned out, the hemp straw bales (made from the stalks of hemp grown for seed production) made excellent building bales, and I went on to use them on many projects.

At that same time, the Internet was starting to play an important role in my sustainable building research, and high on the list of interesting information I was finding online was a material called "hempcrete." All of the early information I could find about hempcrete was from France, and from what I could discern across the language barrier, this material was a mixture of chopped hemp stalks and lime. The recipes were all proprietary, so there was little else to discern other than the basic materials.

Armed with this minimal amount of information, Grant and I began chopping up his hemp stalks and playing with hempcrete recipes. Right from the very first experiments, I was excited and enthusiastic about this material. Our results provided a material that was relatively low-density and would obviously have good insulation properties. At the same time, the material had a body and integrity to it that allowed it to be formed into bricks and blocks with some structural properties. The material could be sculpted and formed. It made an excellent substrate for plaster. I was intrigued.

Early on, I started burying some of the sample blocks to see how they would react to constant exposure to moisture. After one year, I dug out the first buried samples and was impressed to find that there had been no noticeable deterioration. I put them back in the ground, and checked them again at years two and three with the same result, suggesting that this bio-fiber insulation provides resistance to moisture in a way that far exceeds other bio-fiber insulations — very important in the northern climate where I build.

I first used hempcrete in a code-approved building in 2005, casting hempcrete insulated window header sections in a straw bale building and creating one interior infill wall. By 2008, I was using hempcrete to cast fully insulated frame walls, to create sub-slab insulation and to insulate around windows. Many subsequent projects included hempcrete in one form or another.

There was a hiatus in my use of hempcrete for a few years, because Grant decided to stop growing hemp. After a brief period of enthusiastic adoption by farmers in Ontario, the lack of markets for hemp materials in our region led many farmers to cease production. Unfortunately, this situation exists today. It has meant that an important and valuable regional building material cannot be acquired locally. I look forward to the day when this situation is reversed and local farmers once again plant, cultivate, process and sell hemp products locally.

Hempcrete has remained an important part of my building practice, despite having to import the hemp material from another region of Canada. As the European experience has shown, hempcrete insulation has the potential to take an important role in greening our built environment. Academic and government interest in hemp building products has created a healthy market for European hempcrete in just a decade. Since the early work documented in the first book about hempcrete, *Building with Hemp* by Steve Allin in 2005, the number of hempcrete buildings in the UK, France, Germany and other European countries has grown exponentially. Meanwhile, in North America only a small handful of dedicated builders work on developing methods and materials here.

The main reason for the large gap in hempcrete development between Europe and North America is the continued ban on industrial hemp farming in the U.S. The failure of legalized hemp farming in Canada to result in the creation of significant processing facilities shows that without legalization and the development of markets in the U.S., any use of hemp materials on this continent will remain a fringe activity. At the time of writing, there are some positive indications that American federal and state government are moving to relax restrictions on industrial hemp farming. It is now imaginable that a critical mass of farmers and production facilities in the U.S. is possible within a matter of years.

This book is a hopeful precursor to the wide availability of hempcrete materials in North America. The materials and techniques presented here are even now feasible, affordable and practical. The creation of new material sources will help hempcrete in North America follow the positive trajectory of the material in Europe. I encourage forward-looking green builders to jump on board now!

Chapter 2

Rationale

Hempcrete (or *hemp-lime,* as it's commonly called in Europe) is a promising building insulation material. It is also the subject of more hype and hyperbole than any other sustainable building material. Proponents of hemp-based products tend toward unsupported or exaggerated claims of performance and planetary benefit with Websites that make the material seem miraculous.

In truth, hempcrete is simply a very good building insulation material, and there are plenty of compelling reasons to consider using it. It makes an excellent addition to the sustainable builder's "tool kit" of more people- and planet-friendly building solutions. Hempcrete alone will not save the planet, but it will provide an excellent insulation material to a project with the right criteria and context. This book is intended to highlight both the advantages *and* disadvantages of hempcrete and provide potential users with reliable and tested information.

What Is Hempcrete?

Hempcrete is a unique building material, being a *composite* of a bio-fiber (hemp *hurd* or *shiv*) and a mineral binder (lime). These ingredients are blended together with water, and the moistened binder coats all the particles of hemp shiv. A chemical reaction occurs between the lime binder and the water, resulting in the binder setting and gluing the hurd particles together. Generically, it could be called "bonded cellulose insulation."

When the binder is set and cured and any additional water has dried out of the mixture, the resulting material is hempcrete. Unlike many construction composites (such as concrete, mortar and plaster), the binder portion in hempcrete is not intended to fill all the voids between the hemp particles, but only to coat the particles and cause them to adhere to one

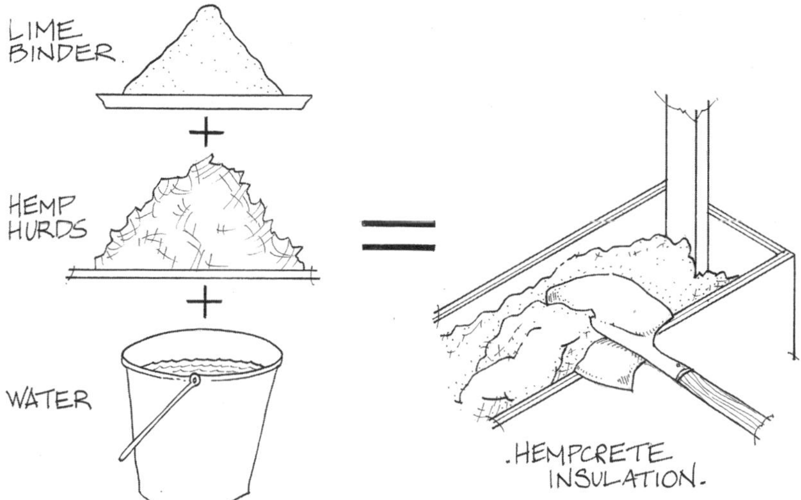

The void space in a hempcrete mix creates pockets of trapped air and is part of the reason the material has desirable thermal properties.

Hemp with lots of fiber.

Hemp with a bit of fiber.

Hemp with some fiber.

The fiber content of the hemp hurd will have a large impact on the qualities of the final mixture. A high percentage of fiber increases density and strength at the expense of thermal performance. Most hempcrete mixtures use little or no fiber.

another where they touch. A hempcrete mix typically has a high percentage of void space in the final mixture.

Hempcrete has a range of desirable thermal, structural and moisture-handling properties that make for an excellent building insulation material. Depending on the mix variables described in this book, hempcrete can be used as roof, wall and/or slab insulation.

Accounting for the variables

Any discussion of hempcrete is complicated by the fact that all three elements in the bio-composite can have a range of types and characteristics, and can be added to the mix in varying ratios.

Hemp hurd

The hemp hurd is the woody core of the hemp plant. It is typically sourced from hemp fiber producers after the valuable hemp fiber has been stripped from the outside of the hemp stalk, leaving the hurd as a by-product.

Variables in the hurd portion of the hempcrete composite include size and grading of the hurd and volume and length of hemp fiber.

Desirable properties for hurd are given in the Material Specifications chapter of this book.

Lime binder

Lime has been an important binder in construction for thousands of years, largely in mortar and plaster recipes. There are different types and grades of lime, and there are various additives that may be included in the binder ratios. These variables will affect the setting time, strength and durability of the hempcrete mixture. There are several brands of manufactured lime binders made specifically for use in hempcrete, and there are recipes for creating lime binder from separate ingredients. The specific qualities that are ideal for a lime binder for hempcrete are discussed in the Material Specifications chapter.

Water

The volume of water added to a hempcrete mixture will have a dramatic effect on the results, even if the hurd and lime variables are controlled. Hemp hurd is extremely porous, capable of absorbing a volume of water that is much larger than what is required for a good hempcrete mixture. Too much water in a mixture can result in higher density, issues with setting for the lime binder, and excessively long drying times for the hempcrete.

Placement and tamping

The final — and also critical — variable is the placement of hempcrete into the building. In wall and floor applications, the material requires some manual tamping in order to ensure that the mix is well bonded and has integrity. However, the amount of tamping can have a major impact on the density of the final material, even if all the mix ratios are identical. Variations in density can have a large impact on thermal performance, so this is an important variable to try to control. This will be covered in the Construction Procedure chapter of this book.

Limitations in quantifying hempcrete

A great deal of this book is intended to help builders find the ideal ingredients and mixtures in order to be able to use hempcrete successfully. But the variability in the formulations and ingredients make it difficult to make blanket generalizations about the performance parameters of hempcrete, especially compared to manufactured insulation products that come from a factory with little or no variation.

Throughout this book, a concerted effort has been made to use relevant testing data, based on knowledge of the variables, and pointing out where the testing data is incomplete or inconclusive.

Why Hempcrete?

The building industry does not see very many "new" materials. The materials used to insulate residential and commercial buildings have remained the same for decades, and most have serious environmental and/or health impacts. As we collectively begin to add more insulation to buildings to lower their energy requirements, the volume of insulating material we use is going to rise dramatically. It makes ecological and financial sense to fill this volume with materials that are annually renewable, low-impact and, ideally, sourced from waste streams or from by-products from other processes. Hempcrete meets all of these important criteria, and compares favorably with conventional insulation materials in many ways.

Affordable insulation

The ingredients for making hempcrete are not common building materials, and as such they do not benefit from the volume price breaks of other insulation options. Still, even prior to wide market availability and the cost reductions this will bring, the cost of hempcrete is comparable

Table 2.1

Material	Approximate cost of 1 square foot of wall surface area @ ~R-28*	Thickness of insulation
Hempcrete	\$3.50 @ \$3.00/ft³ \$11.66 @ \$10.00/ft³	14 inches
Mineral wool batt	\$1.40	7 inches
Fiberglass batt	\$1.20	8.5 inches
Denim batt	\$1.80	8 inches
Dense packed cellulose	\$1.45	9 inches
Extruded polystyrene foam	\$4.40	5.75 inches
Expanded polystyrene foam	\$4.20	7.5 inches
	* cost averages from retail building supply Websites, 2015	

with other insulation options, while bringing advantages over those options in other ways.

The variability in hempcrete pricing is reflected in the chart below, showing that proper attention must be paid to sourcing affordable materials. Hemp sourced from Canadian producers is considerably less expensive than that imported from Europe.

Excellent moisture handling and resistance

Hempcrete is unique among the plant-fiber insulation materials (cellulose, wood fiber, straw bale, straw/clay, cotton) in its ability to maintain integrity in humid conditions. Like all of the plant-fiber insulation options, hemp hurds are able to store a great deal of moisture because of their porous structure; the moisture is *ad*sorbed onto the large internal surface area of the plant fibers and *ab*sorbed into the cellular structure. This storage capacity is very helpful in allowing the material to take on moisture when it exists and to release it when conditions allow. A study performed in France[1] found that up to 596 kilograms (1314 pounds) of water vapor could be stored in 1 cubic meter (35.3 ft^3) of hempcrete, providing storage capacity for a sustained elevated relative humidity of 93% without overwhelming the capacity of the material to adsorb moisture.

The advantage of hempcrete over other plant-fiber materials and conventional insulation types is found in the properties of the lime binder. Lime has a high pH and is inherently antimicrobial and antifungal, and the lime coating around each piece of hemp hurd in the mix creates a surface that resists the development of mold even when the humidity and temperature conditions would cause mold to occur on other insulation materials. This resilience in the presence of humidity or even liquid moisture makes hempcrete unique among insulation materials and a desirable choice in both cold and hot climates and anywhere where humidity levels are high.

Good structural qualities

Hempcrete has a density that allows it to play a minor structural role in the building — unlike batt, loose fill and spray insulation materials in the cost chart above. Hempcrete insulation does *not* have the structural capacity to fully support roof loads, but cast around conventional wall framing or double-stud framing, it can help restrain the studs from bending or buckling under loads, thereby increasing the load that can be carried by each framing member. Testing at Queen's University in Canada showed that a 2×6 wood stud with 313 kg/m^3 (19.5 lb/ft^3) hempcrete infill could support three to four times the compressive loading of a standard stud wall due to the support the hempcrete provides to the wood stud in weak axis bending.[2]

The rigidity of hempcrete insulation and the textured surface it presents on the face of the wall makes an excellent substrate for plaster finishes without any need for mesh or other bonding agents.

An agricultural by-product

Hemp is an agricultural crop that has particularly high yields. A study by the US Department of Agriculture found worldwide yields ranged between 2.5 to 8.7 tons of dry straw per acre.[3] This compares favorably to yields for wheat straw of 1.25 to 2.5 tons per acre.[4] In terms of the amount of biomass available for use from a single crop, no other plant provides as much volume as hemp.

The hemp plant is typically grown for either the strong fiber it produces or for seed (rarely for both at the same time). In either type of hemp production, the hurd is not the primary use and is considered a by-product. It has some market value as animal bedding and can be compressed into fuel pellets, but large-scale hemp production can generate tons of hurd for the insulation market as producers supply fiber or seed to their primary markets.

Good carbon sequestration

According to a 2003 study, 716.6 pounds (325 kg) of CO_2 are stored in one tonne of dried hemp[5]. Tradical, a hempcrete manufacturer in the UK, cites a study showing that their hempcrete product sequesters 110 kg of CO_2 for every cubic meter of material (6.88 pounds per cubic foot)[6] when the carbon emissions from producing the lime binder are taken into account.

In Canada, about 200,000 new homes are built each year, with an average footprint of 2,000 square feet (185 m^2). If they were all insulated to code minimum requirements with hempcrete, a total of 990,718 tons of carbon could be sequestered annually. If the same homes had walls with fiberglass insulation, 207,345 tons of carbon would be *emitted*[7] to create that insulation, so the total net carbon savings for the planet is significant.

Nontoxic building material

Hempcrete is quite a benign material. The farming process uses far fewer pesticides and herbicides than other grain or fiber crops,[8] creating much less environmental damage due to the use of toxins on the fields. The crop does, however, require liberal use of fertilizer, which can have negative ecosystem impacts. Harvesting and processing take place without the input of heat or chemicals.

The dry, powdered lime binder can generate a lot of dust during mixing, and is highly caustic. Adequate breathing protection must be worn by anybody handling the dry ingredients and working around the mixing station. When wet, the lime binder is mildly caustic to skin, so rubber gloves and fully covered skin are required.

Once placed in the wall and fully cured and dry, hempcrete does not off gas or release any toxins into the indoor environment. The lime is antimicrobial and antifungal, and the material is generally thought to have no ill effects on the indoor environment. The excellent moisture-handling abilities of the material can reduce

Table 2.2

Material	Embodied carbon by weight*	Embodied carbon for 4x8 foot wall @ R-28**	Carbon footprint after sequestration
Hempcrete	-2.73 $kgCO_2e/kg$ for 300 kg/m^3 mix	-121.4 $kgCO_2e$	-121.4 kg per 4x8 wall area
Mineral wool batt	1.28 $kgCO_2e/kg$	21.75 $kgCO_2e$	21.75 kg per 4x8 wall area
Fiberglass batt	1.35 $kgCO_2e/kg$	17.6 $kgCO_2e$	17.6 kg per 4x8 wall area
Denim batt	1.28 $kgCO_2e/kg$	22.45 $kgCO_2e$	-1.5 kg per 4x8 wall area
Dense packed cellulose	0.63 $kgCO_2e/kg$	35.3 $kgCO_2e$	-41.3 kg per 4x8 wall area
Extruded polystyrene foam	3.42 $kgCO_2e/kg$	38.5 $kgCO_2e$	38.5 kg per 4x8 wall area
Expanded polystyrene foam	3.29 $kgCO_2e/kg$	37.25 $kgCO_2e$	37.25 kg per 4x8 wall area
	* figures from Inventory of Carbon and Energy (ICE) 2.0	**material densities from *Making Better Buildings*	

the chances of a poor indoor environment due to excessively moist or dry air in the building.

Good, but not exceptional, thermal performance

Hempcrete is an insulation material, and as such its thermal performance is important. One of the primary difficulties in introducing hempcrete to the building industry is the vagary that currently exists around quantifying the thermal performance values of the material. A thorough literature review turns up 19 thermal tests on hempcrete conducted at research facilities around the world. The insulation ratings found by these tests vary widely, from R-1.25 per inch to R-2.3 per inch for low- (200 kg/m^3) to medium- (400 kg/m^3) density wall insulation mixes. Even mixes with the same density vary in the test results. To meet minimum code requirements of R-24 in much of Canada, these results could make the difference between needing a wall that is over 19 inches deep (488 mm) to one that is a more reasonable 10.5 inches (266 mm).

To compound the issue, several *in situ* tests have shown that the actual thermal performance of hempcrete walls is considerably better than the R-values would indicate. Hempcrete has some properties that are unique among insulation materials. As one very comprehensive test of hygrothermal properties of hempcrete states: "The reasonably low thermal conductivity of hemp lime, combined with phase shift, phase change effects, high internal thermal comfort, low initial energy transfer rates, passive humidity control and lower energy requirement for ventilation, all contribute to the reduction of [energy use]."[9] These aspects of hempcrete thermal performance will be explored in the Building Science Notes chapter of this book.

An average of all the test results of mixes in the 275 to 350 kg/m^3 range gives a value of R-1.9 per inch (requiring a 12.5-inch wall to reach R-24). The only North American tests performed to date have been at Ryerson University, and are summarized in the table shown here.

We build our hempcrete walls in the range of 12 to 16 inches (300 to 400 mm) wide for

Table 2.3

Mix and Density	Samples	"k" (W/mK)	"k"- Mean (W/mK) (± 5.00%)	Sample Thickness Mean		RSI (m²K/W)	RSI Mean (m²K/W)	R Imperial Mean	R/Inch Imperial (± 5.00%)
				m	inch				
1 (233 kg/m³)	1	0.075	0.074	0.082	3.21	1.06	1.10	6.23	1.94
	2	0.073				1.13			
	3	0.075				1.10			
2 (317 kg/m³)	1	0.088	0.088	0.081	3.20	0.91	0.92	5.24	1.64
	2	0.088				0.92			
	3	0.088				0.94			
3 (388 kg/m³)	1	0.104	0.103	0.082	3.22	0.79	0.80	4.53	1.41
	2	0.106				0.77			
	3	0.098				0.83			

Source: Dhakal, Ujwal (2016) "The effect of different mix proportions on the hygrothermal performance of hempcrete in the Canadian context," graduate research project at Ryerson University, Toronto.

northern climate use (climate zone 6), and achieve actual performance results that exceed code expectations.

Fire resistance

Although the Internet is full of videos of people aiming blowtorches at hempcrete samples, there is not a great deal of certified testing done on the fire resistance of hempcrete walls. As the homemade videos indicate, the mineral coating of the lime binder around each piece of hemp hurd adds a high degree of flame resistance to the plant material.

A 2009 fire test was conducted by BRE Global in the UK to meet the BS EN1365–1:1999 standard.[10] This test subjected a 3×3 meter (10×10 feet) wall of hempcrete that was 300 mm thick (12 inches) to temperatures of 800 to 1,000° Celsius (1800° Fahrenheit), while also subjecting it to a vertical load of 135kN (30,349 lb). The test showed that the wall met all requirements for integrity, insulation and loadbearing capacity for 73 minutes. During this test, the mean temperature on the side of the wall unexposed to the flames stayed under 60°C (140°F), and for the first 15 minutes stayed under 30°C (86°F). This test was performed on a hempcrete wall with no plaster or other finish on the insulation, so real-world performance would be enhanced by protective plaster or other wallboard.

A Canadian fire test was undertaken in 2015 to ASTM E119-14 and CAN/ULC S101-07 standards. The test report concludes that "The test specimen successfully met the conditions of acceptance for a 68-minute Fire Resistance rating," and included a successful hose stream test.[12]

It is worth noting that the lack of chemical content in hempcrete means that the small amount of smoke generated has none of the highly toxic compounds generated when petrochemical wall insulation and components burn.

Good acoustic properties

There has not been a great deal of testing of hempcrete's ability to dampen sound transmission from outside the house or between rooms. In 2002, a test in the UK was performed on a pair of 6-inch (150 mm) walls with a 3-inch (75 mm) cavity between them, which is a standard arrangement for walls separating units within a building. The hempcrete walls offered sound reduction of 57 to 58 dB, exceeding the 53 dB code requirement.[11]

Who Would Want to Build with Hempcrete?

Currently, hempcrete is a material choice for owners and builders who wish to create a building with the following qualities:

- Low- or zero-carbon footprint
- Nontoxic materials and high indoor air quality
- Excellent moisture-handling properties
- Durability
- Fire resistance
- Good thermal performance and very stable indoor temperatures

In exchange for these positive qualities, the builder will face slightly higher initial material costs and additional effort to source the materials from non-standard channels.

Current trends in North America are toward carbon reductions, less toxic materials, moisture resilience and durability; a builder who establishes an early foothold in the hempcrete market is likely to be rewarded as more people make choices based on the qualities that hempcrete has to offer.

One of the key advantages of hempcrete over other natural insulation materials is that it fits well with conventional framing techniques, so although it requires a different installation process than conventional insulation, it does not necessarily require experienced builders to change their approach to framing. And the formwork used for installing hempcrete is a lightweight version of the formwork familiar to any builder who has worked with poured concrete. Forming hempcrete in a frame-walled building results in perfectly straight and flat walls, so that the aesthetic result also closely matches mainstream expectations.

In particular, hempcrete is a good choice for those who live in extreme climates, either cold or hot, and in places where humidity levels are high for sustained periods of time. The antimicrobial and antifungal qualities of hempcrete make it very stable and safe when highly loaded with moisture, and can help to prevent mold and deterioration in such conditions.

There is an underlying tension among hempcrete advocates between those with a do-it-yourself philosophy and those aligned with the companies that have proprietary formulations for hempcrete. There are some real advantages to working with materials that have been tested and developed by a company with some history and knowledge and customer support. At the same time, the materials are simple, straightforward and accessible to those who have a desire to formulate and mix their own hempcrete. There is no right or wrong tactic to take, and it is the aim of this book to cover both options fairly and thoroughly.

As the hemp industry begins to take hold in North America, there will be improved supply chains and reductions in cost that will make hempcrete even more attractive. Early adopters using the material today are helping to set the stage for an increase in the use of hempcrete in the near future.

Notes

1. Evrard, Arnaud and De Herde, André. "Bioclimatic envelopes made of lime and hemp concrete." CISBAT 2005 *Renewables in a Changing Climate: Innovation in Building Envelopes and Environmental Systems* (EPFL, Lausanne, Switzerland, 09/28/2005 to 09/29/2005).
2. "Structural benefits of hempcrete infill in timber stud walls," Agnita Mukherjee. Thesis submitted to the Department of Civil Engineering, Queen's University, Kingston, Ontario, January 2012.
3. "Industrial hemp in the United States: Status and market potential," Economic Research Service. Staff Report No. (AGES-001E), January 2000.
4. "Straw production and grain yield relationships in winter wheat," Edwin Donaldson, William Schillinger, and Steve Dofing. Pacific Northwest Conservation Tillage Handbook Series No. 21, June 2000.
5. Pervais, M. 2003, "Carbon storage potential in natural fiber composites," *Resources, Conservation and Recycling,* vol. 39, no. 4, pp. 325–340.
6. tradical.com/pdf/Tradical_Information_Pack.pdf.
7. Based on 1.35 kgCO2/kg from "Inventory of Carbon and Energy 2.0," Geoff Hammond and Craig Jones. University of Bath, 2011.
8. "Industrial hemp in the United States."
9. Lawrence, M. et al. "Hygrothermal performance of an experimental hemp-lime building," *Key Engineering Materials,* 517, pp. 413–421, 2012.
10. BRE Global, test report #250990.
11. "Final report on the construction of the hemp houses at Haverhill, Suffolk," Client report number 209–717 Rev.1. BRE 2002.
12. QAI Laboratories, Test Report #T1007–1, June 8, 2015, Coquitlam, B.C.

Chapter 3

Appropriate Use

General Use Parameters

Hempcrete insulation has four common uses in buildings: as wall insulation, roof insulation, sub-slab insulation, and window insulation and trim. Each is discussed here, and further details for mixing and applying each of these will be given in subsequent chapters.

Wall insulation

Hempcrete of a density ranging from 18.75 to 25 pounds/ft^3 (250 to 350 kg/m^3) can be used to insulate the exterior walls in most low-rise (three story or less) construction scenarios described in the International Residential Code (US) or in Part 9 of the National Building Code (Canada). The walls will require structural components — most commonly a wooden frame — that meet the requirements of the code, and the hempcrete insulation will be an infill within the wall structure or formed on the exterior or interior side of the structural frame.

An engineer can use the results of a wide range of strength testing data (see Resources) to determine the degree to which a structural frame may be supported by the hempcrete wall insulation. Stud dimensions may be able to be

Hempcrete walls are regularly framed with 2×4s at 24-inch spacing instead of 2×6s at 16-inch, and it is possible to decrease wood use further with an engineer's assistance.

Marks & Spencer store.

Hempcrete is laid untamped on the flat attic ceiling. CREDIT: BOB CLAYTON

Hempcrete insulation can be used beneath floor slabs if the loads are calculated to be suitable. A vapor barrier between the ground and the insulation prevents soil moisture from saturating the insulation.

reduced and/or the spacing between studs may be able to be increased to save on lumber.

Whether hempcrete is suitable as a wall insulation for larger projects and/or projects covered by other code sections must be determined by appropriate design professionals. The panels have been used as curtain walls in larger projects, up to six stories in height.

Roof insulation

Hempcrete of a density ranging from 12.5 to 15.6 pounds/ft³ (200 to 250 kg/m³) can be used to insulate the roof of most buildings. The material can be used as loose-fill insulation on flat ceilings or it can be lightly tamped into vaulted roof assemblies.

As hempcrete is heavier than most conventional roof insulation materials, appropriate calculations must be made to ensure that the roof framing and ceiling materials are designed to carry the additional loads.

As roof insulation, hempcrete has the advantage of discouraging pests, and it will not blow or settle over time. The material is also much more resistant to moisture than conventional insulations. Given that most buildings experience several roof leaks over their lifespan, a moisture-resistant insulation can provide important resilience.

Sub-slab insulation

Less common than wall and roof applications, hempcrete of a density ranging from 23.4 to 31.2 pounds/ft³ (375 to 500 kg/m³) can be used to provide insulation underneath floor slabs. Load-bearing capacity can range from 160 to 290 pounds per square inch (1.1 to 2.0 Mpa), which is suitable to support many floor slabs. The higher density for slab insulation lowers the thermal value, with ranges from R-1 to 1.5 per inch. Under-floor insulation typically requires lower R-values than wall and roof insulation, due to the warmer and more consistent temperature difference between the earth and the building, compared to the building and outdoor air temperatures.

Hempcrete used under a floor should be placed on a stable, well-drained base and a vapor barrier should be employed between the soil and the insulation.

Window insulation and trim

This use for hempcrete is not widespread, but the author has had numerous successful applications and sees this as a potentially valuable use for the material. A hempcrete mix that includes a percentage of hemp fiber is used to fill the shimming space between window units and framing, replacing stuffed batt insulation or expanding spray foam. The window can then be trimmed conventionally, or the hempcrete can be formed to create an exterior and/or interior trim for the window. This creates a unique aesthetic opportunity and a means of providing an airtight seal around windows without any reliance on petrochemical products.

Other forms of hempcrete

Precast formats: Some experimentation is being done in Europe with making precast hempcrete panels and blocks. Made and dried in a production facility, the cured panels or blocks are transported to the site and installed.

Hempcrete blocks have the disadvantage of requiring some form of mortar to glue the

The sculptable nature of hempcrete mixed with some hemp fiber makes it possible to form window trim while using hempcrete to insulate around windows.

Two experimental forms of hempcrete are combined in this block cast using clay slip as the binder instead of lime.

blocks together, adding to site installation time and creating thermal bridges where the mortar provides lower thermal performance across the wall than the hempcrete. Interlocking blocks have been used experimentally to reduce or eliminate the need for a full-width mortar joint.

Hempcrete panels have seen some limited use in the UK. The hempcrete is formed into a wooden frame, and the frame and insulation are installed to the exterior of the building, usually wrapping a structural post and beam frame.

Precasting the hempcrete eliminates the drying/curing time from the build schedule, as the materials arrive ready to be plastered or clad as soon as they are installed. Both forms of precast hempcrete have significant potential to help bring hempcrete into more mainstream construction applications. Hopefully, North American entrepreneurs will begin to develop the material in these forms.

Clay/hemp: In modern natural building, there are many examples of clay being used to bond cellulose fibers like straw and woodchips (known as light straw clay, slipstraw or EcoNest™ construction). Clay could potentially replace lime as a binder with hemp hurds, further reducing the environmental impact of the material. Many properties of the material would change with clay as the binder, including longer drying times and different structural and thermal properties, but it's certainly a feasible option worthy of exploration.

Chapter 4

Building Science Notes

THE PRINCIPLES OF BUILDING SCIENCE can and should be applied to all elements of a building enclosure throughout the design and construction phases. Managing the flow of heat, air and moisture as they relate to insulation materials is the key to creating durable, healthy, high performance buildings. Viewed through this lens, the functionality of insulation materials are affected by four different control layers:

- Thermal control
- Air control
- Vapor control
- Water control

Hempcrete can be part of a well-designed building that adheres to all of the principles of building science.

Thermal Control: Principles

The movement of heat is always from an area of higher concentration to an area of lower concentration (from warmer to colder). There are three different modes of heat transfer: radiation, conduction and convection.

Radiation: Radiant heat energy comes from any warm mass, including the sun, heating devices, human bodies, or other warm objects. Cooler materials struck by this radiation absorb the heat energy. Heat transfer by radiation is controlled when the path of the heat radiation is blocked. If you can't see daylight through a material, it does a good job of controlling radiant heat transfer because the material blocking its path absorbs the heat energy.

Conduction: Conductive heat energy moves by direct contact between materials, as warmer areas transfer heat to cooler areas. Heat transfer by conduction is controlled when the material being warmed has poor conductivity. The less dense a material, the poorer the thermal conductivity, the better for controlling conductive heat transfer.

Convection: Convective heat movement is provided by the movement of warm fluid (for buildings, the main "fluid" is air), which becomes less dense and rises as it gets warmer, and becomes denser and falls, as it gets cooler. Convective heat transfer is controlled in building assemblies by reducing the volume of air that is free to circulate convectively. This is why voids and gaps in insulation can have a dramatic effect on thermal performance, and why it is important to have insulation materials making consistent contact with adjacent layers.

The thermal control layer — insulation — cannot prevent the movement of heat, but it can effectively slow all three forms of transfer through the assembly in both directions in order to preserve a comfortable interior temperature using the minimum amount of heating or cooling input energy.

For a given climate, the thermal control layer must meet a standard of thermal control required by building codes or voluntary energy efficiency standards. The effectiveness of this thermal control layer is typically measured by the static-state conductivity (U-value) or resistance (R-value). However, testing for thermal resistance at steady-state temperatures and in a moderate temperature range does not always reflect real-world dynamic performance. Although not conventionally considered when assessing thermal insulation, thermal diffusivity and thermal effusivity are important elements in understanding the effective thermal control of a building.

Any breaks in the thermal control layer can cause degradation of performance. These "thermal bridges" are caused by voids (gaps, cracks and settling) in the insulation or by materials within the assembly (most typically, framing members) that have a poorer thermal resistance than the insulation, providing a "short-circuit" for the flow of heat by conduction through the wall.

Thermal control: Application for hempcrete

The issue of assessing the thermal control properties of hempcrete is very important for the widespread adoption of hempcrete in North America, and unfortunately, it is not a straightforward subject. The problem is that measured steady-state R-values for hempcrete are quite low compared to conventional insulation materials, as shown in the testing ranges in the table shown here.

Even the best results for hempcrete are far from the R-3.5 per inch of dense packed cellulose insulation, or other conventional options. Hempcrete has some definite advantages over other insulation in other aspects of building science, but judged strictly by static-state R-value figures, it does not seem to measure up.

However, building researchers are searching for new metrics to measure and compare thermal performance that do not rely on steady-state R-values, and this work will perhaps put hempcrete on a scale that is more indicative of its real-world performance. Several studies examining the hygrothermal performance of hempcrete have shown that actual performance of hempcrete insulation far exceeds the predictions made by R-value. Using computer simulation models that take hygrothermal properties into account, one test found "differences of ... 19.5% in energy consumption,"[1] which is a significant improvement over the predicted energy use. The same study examined cellular concrete, which, at a density of 30 lb/ft^3 (480 kg/m^3), has the same R-value as hempcrete of the same density at R-1 per inch. "[B]ehaviour was compared to that of cellular concrete and it was found that reduction of 45% in energy consumption can be reached." This type of modeling also correlates with real-world measured experience. In a study of a public housing project in Suffolk, England, two hempcrete units were built to an identical specification as two brick cavity homes for comparison purposes.[2] Using static-state R-values, the conventional home with rock wool insulation had a SAP[3] rating of 87, versus a much lower SAP of 77 for the hempcrete house. However, one unoccupied unit (removing occupant differences) of each type of house used exactly the same amount of heating energy to maintain the same indoor temperature, which undermines the assumption that the hempcrete would perform worse because of its lower R-value. In the occupied houses, the hempcrete home used four percent less energy, despite the fact that "the temperatures maintained in the

Table 4.1

Density of hempcrete	R-value per inch range*	Thermal conductivity range**
Lightweight mix: 150 to 250 kg/m^3 9.4 to 15.6 lb/ft^3	R-2.4 to R-1.8 per inch	0.06 to 0.08 w/m.K
Medium mix: 250 to 350 kg/m^3 15.6 to 21.8 lb/ft^3	R-1.9 to R-1.44 per inch	0.08 to 0.1 w/m.K
Heavy mix: 350 to 500 kg/m^3 21.8 to 31.2 lb/ft^3	R-1.44 to R-1.03 per inch	0.1 to 0.14 w/m.K

* Calculated from metric thermal conductivity figures, based on a 300 mm (11.8 inch) wall thickness.

** Range of test results based on all available documents. See Resources for list of tests.

hemp houses have been consistently one or two degrees higher than in the brick houses."

Another study used WUFI, a software program for calculating the coupled heat and moisture transfer in building components, to examine heat flow in a hempcrete wall:

> The energy lost from hemp lime in the first 24 hours is 187KJ/m^2, which equates to an average heat loss of 0.11 W/[m^2·K] despite the fact the theoretical U-value for this thickness of hemp lime is 0.29 W/[m^2·K] (Evrard & De Herde, 2005). This is evidence of how dynamic thermal performance can be different from predictions based on steady state figures/U values.
>
> The same simulation provides evidence of the ability of hemp lime to almost completely (98.5%) dampen a sinusoidal change in external temperature of 20°C to 0°C over a 24 hour cycle with a time shift of 15 hours, the time delay of the peak temperature getting through the wall. This compares to a dampening of 77.5% for mineral wool with a time shift of only 6 hours and with a dampening of 95% for cellular concrete with a time delay of 10.5 hours (Evrard & De Herde, 2005).[4]

Claims are often made that the "thermal mass" of hempcrete gives it better thermal performance than other insulation materials. This is not true; greater density results in more conductive heat movement. Thermal mass is not insulation. What sets hempcrete (and other bio-fiber materials) apart is its hygrothermal performance. In making the case for a new kind of assessment of thermal performance for bio-fiber building materials, a report from the University of Bath explains it well:

> These materials are vapour active and their response to changing humidity conditions is associated with their pore structure and pore connectivity. Their adsorption/desorption characteristics involve thermal effects from latent heat to the extent that moisture condenses (releasing heat) and evaporates (absorbing heat) on the surface of the material and within its pores (Hill et al., 2009). This phenomenon increases their effective thermal mass, allowing them to act as a thermal buffer in conjunction with their hygric buffering properties.[5]

This hygrothermal analysis of hempcrete is a long way from being codified, or even recognized, in the building industry. Until then, thermal control should be achieved by at least meeting minimum code standards for insulation by using an appropriate thickness of hempcrete based on tested R-values. By meeting code requirements based on static R-value, the homeowner should benefit from higher than expected thermal performance.

Air Control and Thermal Performance: Principle

The movement of air through a wall assembly "short-circuits" the effect of the insulation, as heat is carried through the wall at an accelerated rate; this drastically lowers the thermal performance, regardless of the R-value of the insulation in the wall. Even at a relatively low pressure difference between inside and outside of 10Pa, heat flow through a wall can be nearly 25–50% higher than what the R-value of the wall would be with no air movement.

An effective air control layer (or layers) must work as a continuous barrier to the free movement of air. Each element of the building enclosure must have an air control layer, and the seams between elements must also be designed and built to prevent air leakage. The continuity of the air control layer is a crucial detail that must be considered at the design stage and integrated into the construction.

Continuity of the air control layer must also be maintained at every penetration through a wall or other element, including electrical outlets, pipes and service conduits.

Air control and thermal performance: Application for hempcrete

A proper installation of hempcrete insulation should be completely free of voids and spaces. The material does an exceptionally good job of preventing the flow of air through the wall even prior to the addition of plaster, sheathing and/or any other air control layer. According to a study at the University of Bath, a hempcrete structure with no formal air control layer had even less air permeability than the value targeted by the stringent Passivhaus specs.[6]

While this degree of air tightness far exceeds current North American code requirements, properly detailing the plaster or sheathing layers on the hempcrete can ensure the air tightness further and make sure that any small leaks caused by defects in the installation do not subvert the overall air tightness.

The use of vapor permeable house wrap air barriers on the exterior of a hempcrete wall is required by codes in many jurisdictions, and can help to ensure a high degree of air tightness. There is no issue with using an exterior house wrap over hempcrete insulation, unless you are planning to use a plaster finish applied directly over the hempcrete. In this case, the plaster will perform as the permeable air barrier. This is a viable solution, but may require you to pursue alternative compliance pathways in the building code to be approved (see the Building Codes and Permits chapter).

As with any building, special attention should be paid to air tightness between walls insulated with hempcrete and junctions with other materials. In particular, the joint between the top of the wall and the ceiling should be protected from air leakage, as this seam will not benefit from the air tightness of the hempcrete insulation.

Air tightness is especially important for hempcrete attic insulation. The lower density of attic mixes may not provide the same degree of air tightness as the denser wall systems that have been lab tested. For this reason, using a distinct air control layer beneath the insulation is important. This layer can be an appropriate sheet-style material, airtight drywall application, plaster or any other material that will prevent the movement of air into the attic. Penetrations through this air control layer (electrical boxes, plumbing pipes) should be carefully detailed to maintain the integrity of the air control.

If you are using hempcrete to insulate around windows and are attempting to create a very airtight building, it is recommended that you use a high performance sealing tape between the framing and the window unit on at least one side of the window. The volume of hempcrete in a window shim space may not be sufficient to provide the degree of air tightness that is achieved from a solid 12-inch (300 mm) thick wall.

Hempcrete used to insulate under a floor slab should be sealed off from the exterior air around the perimeter of the building to prevent air infiltration into the insulation.

Air Control and Moisture Performance: Principle

The movement of air through a wall assembly can carry a significant amount of water vapor into the wall. Warmer air holds a higher quantity of water vapor and will be driven toward areas of cooler, drier air. As this air cools down, its ability to hold water vapor decreases, and it will condense and deposit its moisture load in the wall. This can cause serious issues of mold, rot and deterioration of the wall materials over time.

Relatively small holes through which air can enter a wall assembly can carry large amounts of moisture into the wall. Even a hole as small as 1×1 inch (25×25 mm) can carry as much as 7.5 gallons (28 liters) of moisture into a wall over the length of a heating season (figures for Ottawa, Canada), making air sealing a vital part of controlling moisture migration.

Air control and moisture performance: Application for hempcrete

Controlling moisture that is moving into the wall via air leakage should be handled in the same manner as described in Air Control and Thermal Performance.

If all potential leakage points are addressed and air movement through the insulation is controlled, the moisture being carried into the wall will be reduced to levels that are not of concern for hempcrete insulation.

Vapor Control: Principle

In many ways, the air control layer in a wall system performs the function of a vapor control layer. The vast majority of moisture entering a wall system is carried by free moving air (see above). If air movement is controlled, a small amount of moisture will still enter the wall by diffusion.

In the diffusion process, vapor pressure caused by a difference in moisture content on either side of a wall causes moisture to migrate at a molecular level through pore spaces in the wall materials.

The ability of a material to resist this diffusion of moisture is measured by units called "perms." There are three classes of vapor retarders, based on perm ratings:

- Class I — 0.1 perm or less (qualifies as a vapor barrier, or vapor impermeable)
- Class II — 0.1 to 1.0 perms (vapor semi-impermeable)
- Class III — 1.0 to 10 perms (vapor semi-permeable)

Materials with a perm value higher than 10 perms would be considered completely vapor permeable.

Vapor control: Application for hempcrete

Hemp hurd has a unique pore structure that gives the material an excellent ability to store and release moisture. "Hemp-lime is capable of rapid liquid transfer, high moisture retention and high water vapour permeability, all of which act to avoid condensation, and manage the internal environment to retain comfortable conditions."[7] Unlike conventional insulation materials, the passage of vapor into the insulation is not problematic, as storage capacity is very high and will not be overwhelmed by periods of high vapor drive. Along with storage capacity, the material has high permeability, allowing passage of vapor through the wall to be released to drier air. Maintaining the moisture balance in a wall system is the key to long-term performance and durability.

This strategy of leaving a wall system "vapor open" is unconventional, and is not supported by current codes. There is, however, a growing body of building science research that points to the many advantages of vapor-open wall systems. A wall that can allow moisture to pass all the way through in either direction is much less likely to suffer from moisture overloading issues

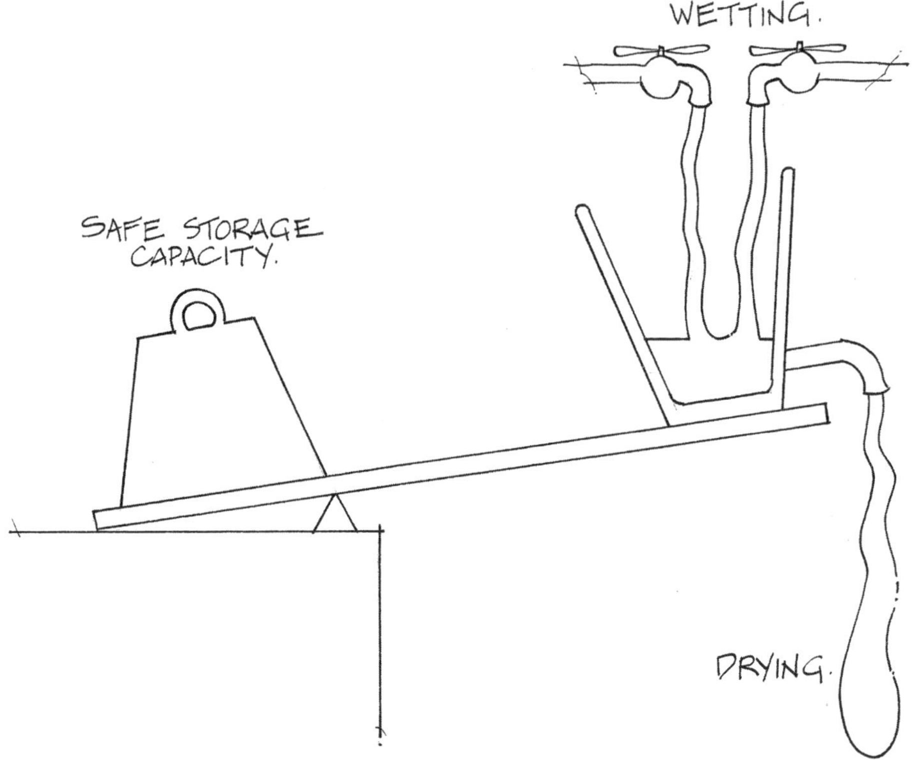

such as wet rot and mold. Vapor-open walls acknowledge that despite a builder's best efforts, moisture is likely to get into the wall at some point during the building's lifespan, and giving it a free and easy path out of the wall is a more resilient strategy.

For hempcrete walls, maintaining vapor permeability in the whole system is critical. This means that any finishes applied to the wall must also be vapor permeable. Materials with a perm rating of 4 (the rating of a 1:1:6 lime:cement:sand plaster[8]) or higher are suitable. An exception to this rule would be for materials/finishes applied over a ventilated cavity.

For hempcrete attic insulation, a Class I or II vapor control layer should be used to reduce migration of moisture into the attic space. The relatively still air in attic spaces has significantly less ability to diffuse moisture than freely moving outside air, and the underside of the roofing can be much colder than the attic air, making condensation on these surfaces much more likely. Codes mandate adequate ventilation of attic spaces to help mitigate this risk, but the degree of ventilation is based on good vapor control and could be overwhelmed by too much migrating moisture.

Hempcrete slab insulation should be protected from rising dampness from the soil by an adequate Class I vapor retarder. Slab insulation has little to no opportunity for drying and therefore must be protected from the intrusion of moisture.

Water Control: Principle

Bulk loading of moisture from rain is the highest risk for walls. The design of a building should consider this risk and mitigate exposure as much as is practically possible. Roof overhangs, foundation toe-ups and proper window flashing and sills are all key elements of water control.

Inevitably, the wall will receive some bulk loading of water on the exterior. The degree of protection required will depend on climate conditions (amount and frequency of rain, strength and direction of winds, drying conditions). Protection can come in the form of surface coating and/or rain-screen protection.

Surface protection comes from the final cladding material applied to the exterior face of the wall, or a combination of the cladding material and a protective coating (typically paint).

A rain screen is made by creating a ventilated space in front of the main air/weather barrier on the exterior face of the wall, and affixing a final cladding material in front of this ventilated space. This allows the vast majority of precipitation to be shed by the cladding, with the ventilation space allowing the cladding and the wall assembly a drying channel that is protected from the elements. This is a more resilient, durable approach than surface protection, but adds another layer of material, effort and cost to the building.

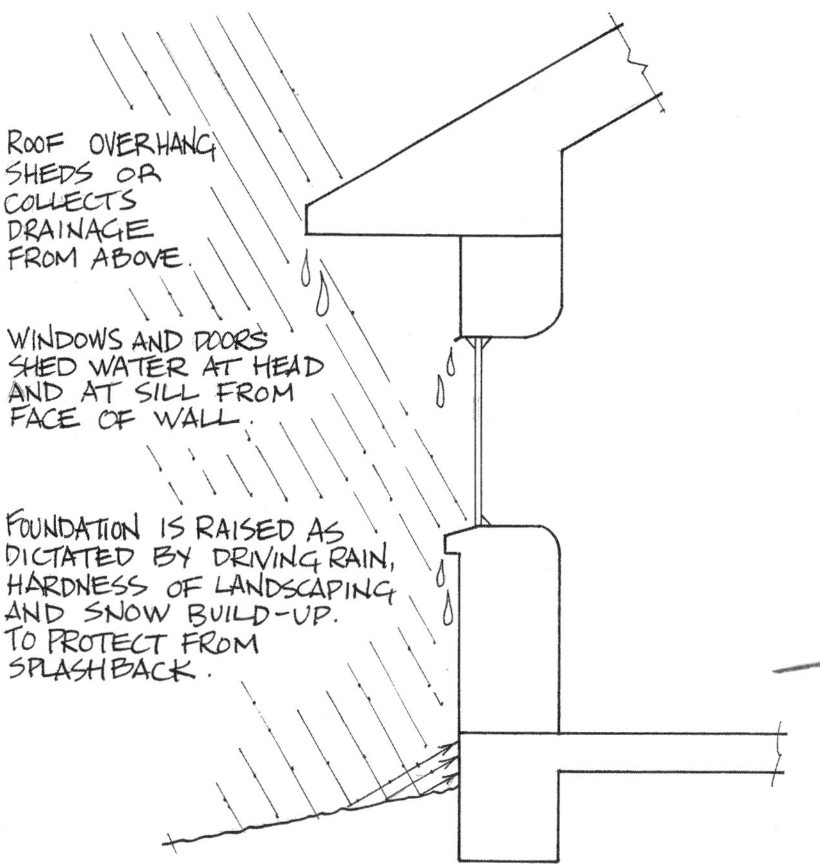

Water control: Application for hempcrete

Protection from precipitation loading for hempcrete walls can come two forms:

Surface protection of plaster: Where plaster applied directly to the hempcrete will be the finished exterior surface, it is possible for the plaster to become super saturated by precipitation and result in moisture loading of the hempcrete behind.

Water spray testing was carried out on hempcrete walls to explore the ability of hempcrete to handle high moisture loads. "[O]ver the spray period of 96 hours the test walls were sprayed with a minimum of one year's wind-driven rain similar to that at a severely exposed location, or 5 years at a less exposed site."[9] The results showed no penetration of water into the hempcrete after two hours, and limited penetration of ¾ to 1½ inches (20 to 40 mm) after 24 hours. After the full 96 hours, water penetration was just 2 to 2¾ inches (50 to 70 mm). Given the ability of hempcrete to dry out quickly and resist the onset of mold or rot (due to the presence of the lime binder), this seems to indicate that direct-plastered hempcrete is sufficient water control in all but the worst climatic conditions.

It is possible to greatly increase the water control properties of plastered hempcrete with the addition of permeable paint on the surface of the plaster. The best option is a water-repellant paint with a Class III or high perm rating (10 perms or more), such as potassium silicate paint. Using a protective coating with a Class I or II perm rating will cause vapor/moisture control issues and must be avoided.

Rain screen: A rain screen can be added to the exterior face of a hempcrete wall, built over a plaster substrate or a structural sheathing material.

Vertical strapping applied to the face of the wall creates open channels that keep moisture-laden cladding from touching the wall and provide a pathway for drying air to carry away moisture from the cladding and the wall. These channels must be protected from intrusion by insects or animals.

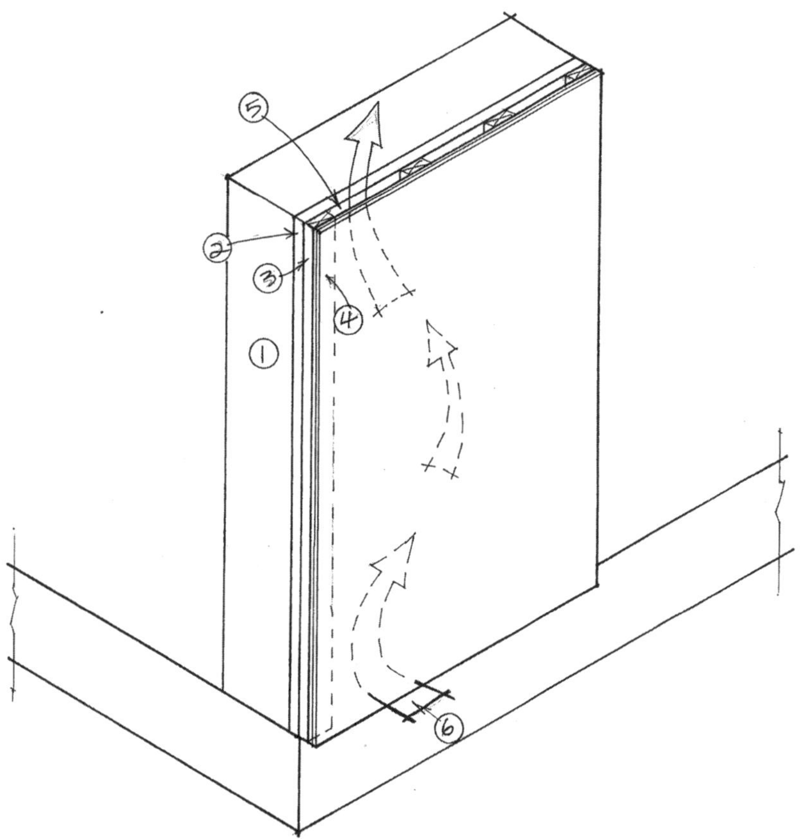

1) Hempcrete insulation
2) Plaster or permeable sheathing
3) Vertical strapping
4) Exterior cladding
5) Ventilation channel
6) Air flow through ventilation channel

For hempcrete attic insulation, bulk water control comes from the ability of the roof to shed water. As long as the roof is in good condition, the hempcrete insulation will be protected.

Notes:

1. Tran Le, A.D. et al. "Transient hygrothermal behaviour of a hemp concrete building envelope," *Energy and Buildings* 42(10) 1797–1806, September, 2010.
2. "Final Report on the Construction of the Hemp Houses at Haverhill, Suffolk." Client report number 209-717 Rev.1. Suffolk Housing Society Ltd, August 2002.
3. Standard Assessment Procedure (SAP) is the methodology used by the UK Government to assess and compare the energy and environmental performance of dwellings.
4. "Hemp lime bio-composite as a building material in Irish construction." Prepared for the Environmental Protection Agency by BESRaC (Built Environment Sustainable Research and Consultancy).
5. Lawrence, M. et al. (2013) "Hygrothermal performance of bio-based insulation materials," *Proceedings of the Institution of Civil Engineers: Construction Materials* 166 (4) pp. 257–263.
6. Shea, A., Lawrence, M. and Walker, P. (2012) "Hygrothermal performance of an experimental hemp-lime building," *Construction and Building Materials* 36, 270–275.
7. Lawrence, M. et al. (2012) "Hygrothermal performance of an experimental hemp-lime building," *Key Engineering Materials* 517, pp. 413–421.
8. *Design of Straw Bale Buildings.* Bruce King et al. Green Building Press, 2006, pg 156.
9. "Final Report on the Construction of the Hemp Houses at Haverhill, Suffolk," Suffolk Housing Society Ltd, 2002.

Chapter 5

Material Specifications

Hemp Hurd

The hurd is the woody core of the hemp plant, and makes up about 60–70% of the volume of the stalk, the other 30–40% being the fiber content. Whether the hemp is being grown for fiber or for seed, the hurd is generally considered a by-product. "The results from chemical analysis show that hemp hurds contain 44.0% alpha-cellulose, 25.0% hemicellulose, and 23.0% lignin as major components, along with 4.0% extractives (oil, proteins, amino acids, pectin) and 1.2% ash."[1]

Specifications

There are no formal specifications for the hemp hurd used for making hempcrete. To date, any standards used by hemp producers to create hurd for the building market are self-imposed, and there are no third-party specifications. Current manufacturers of hemp hurd do not publish their specifications.

There is debate surrounding the inclusion of hemp fiber along with the hurd. My own experience is that too much fiber content (more than 10–20%) makes the mixing process more difficult, slows down drying times and increases density. I have much preferred hurd sources that have little to no fiber content, and all of the hurd packaged specifically for use in hempcrete comes in this form.

Dimensions

Though there are no published standards; on-site measurements of raw hemp hurds purchased for making hempcrete show a particle size ranging from ⁵⁄₆₄ to 1 inch (2 to 25 mm) in length and ³⁄₆₄ to ¹³⁄₆₄ inch (1 to 5 mm) wide.

Ideally, less than 0.5% of the mass should be fines less than ¹⁄₆₄ inch (0.5 mm), and 75–90% of the mass should be between ³⁄₆₄ to ¾ inch (1 to 20 mm) in length. Less than 10–25% should be greater than ¾-inch long (20 mm). The hemp should be free of dust. Fines absorb water and lime binder and upset the mix ratios and the final density of the hempcrete.

The grading and distribution of particle sizes will affect many performance aspects of the hempcrete mix, including strength and thermal properties. A lot of small particles will result in a mix that requires more binder and is denser, giving the composite more strength and less thermal resistance. A lot of large particles will result in a mix that requires less binder and is less dense, giving the composite less strength. A relatively even blend of small, medium and large pieces is ideal.

Density

The average dry density of hemp hurds ranges from 6.8 to 7.2 lb/ft^3 (110 to 115 kg/m^3). Hurd may be compressed into bales or bundles for shipping and will be denser than this range until unpackaged.

Moisture Content

As with all cellulose-based building materials, the moisture content of hemp hurds prior to mixing should be less than 20%.

Sourcing

There is currently only one source for North American hemp hurd graded and packaged for use in hempcrete: Plains Hemp in Manitoba,

Canada. Several hemp distribution companies also import European hemp from the UK, France, The Netherlands and Slovakia. Some Chinese hemp hurd is also available in North America. See the Resources section of this book to find supplier contact information.

Several European companies that sell proprietary hempcrete mixtures sell both the lime binder and the hemp hurd, and insist that customers must use both components, or neither. The advantage of these products is that they come with mix formulas and some degree of technical support and warranty. The drawback is that they can be significantly more expensive than sourcing hemp and lime separately, as both elements must be shipped across the Atlantic.

Cost

North American (currently, Canadian) grown and processed hemp hurd is available for $0.16 to $0.18 USD per pound, and significantly less expensive than that imported from Europe or Asia, which averages $0.22 to $0.63 USD per pound.[2]

The cost of Canadian hemp equates to $1.20 per cubic foot ($42.40 per m^3). Imported hemp ranges from $1.50 to $4.33 per cubic foot ($53.35 to $152.78 per m^3).

Shipping costs will vary widely depending on quantity, distance and mode of shipping. Be sure to get quotes for shipping when pricing hemp hurds from suppliers, as sometimes less expensive hemp comes with more expensive shipping.

Table 5.1: Environmental impacts of hemp hurd

Ecosystem Impacts	Embodied Energy	Carbon Footprint	Indoor Environment	Waste
Low. Hemp production involves the use of little or no pesticides and herbicides, and is often grown on marginal farmland. There are impacts from the use of fertilizers on the crop, with quantities varying by region and farming practices.	Very Low. 1.09-6.19 MJ/kg.[1] There is very little published data on the EE of hemp hurd production, but it is a low input crop and does not require heat or energy intensive processing.	Low. 0.56 kgCO_2/kg. With sequestration of CO_2 within the hurd taken into account, there is a value of -1.27 kgCO_2/kg.[2]	Low. There are no known toxins from hemp hurd, and no off gassing. Dust during mixing should be avoided.	Low. Unused hurd can be stored indefinitely. The material is also fully compostable. Leftover hempcrete can be recycled into a new hempcrete mix.

Notes:

1. "The carbon sequestration potential of hemp-binder: A study of embodied carbon in hemp-binder compared with dry lining solutions for insulating solid walls." Naomi Miskin. MSc. Architecture: Advanced Environmental and Energy Studies, January 2010
2. Ibid.

Lime Binder

There are several options for the composition of the lime binder portion of a hempcrete mix. The advantages and disadvantages of each type of binder will be covered in the chapter on mixes. Here, only the material properties of the ingredients will be considered.

Lime has a long history of use in buildings around the world dating back several thousand years, most commonly as the binder in mortar and plaster. Although not nearly as popular a material as it once was (largely displaced by Portland cement), it is still available, and there are standards in place for most codes to cover its use in mortar and plaster. No standards or common specifications exist for the use of lime in hempcrete.

In considering lime binders, it is important to understand the difference between the two kinds of lime used in construction: *hydrated lime* and *hydraulic lime.*

Hydrated lime cures chemically by re-absorbing the carbon dioxide that was driven out of the limestone during production. In order for this process to happen properly, the lime must have contact with the atmosphere in order to have access to free CO_2. This carbonization process takes place over a long period of time, and hydrated lime plaster and mortar gains strength over months, years and even decades, as does hempcrete made with hydrated lime. In general, hydrated lime should not be used as the sole binder for hempcrete because the lime in the middle sections of thick layers of insulation will not sufficiently recarbonize and so will not reach their full working strength. This may or may not be critical, but it is the reason that most hempcrete binder formulations include some form of additive to produce a hydraulic reaction, in which the chemical set of the material happens in the presence of water.

Hydrated lime is produced in North America and is widely available through masonry supply outlets in North America. Type S lime is the most commonly available form of hydrated lime.

Hydraulic lime cures chemically as well, partly by reacting with water to create a new compound, and partly by carbonizing. The hydraulic set of this type of lime means that thick layers of hempcrete will reach a reasonable portion of the working strength of the lime fairly quickly. A natural hydraulic lime with a hydraulic reaction of 5 is most commonly used for hempcrete.

Hydraulic lime is not produced in North America. There are distribution networks established for imported European natural hydraulic lime. In larger cities, masonry supply outlets may stock hydraulic lime, or it may require special ordering through smaller masonry supply centers.

Natural cement binders are sometimes used in hempcrete mixes. One North American brand, Rosendale Cement, is still being produced, and there are also imported versions from the UK and Europe. Natural cement is similar to hydraulic lime, but is made from argillaceous limestone and is not slaked after being fired, but ground into a powder that has a quick hydraulic set. Natural cement can take the place of natural hydraulic lime in hempcrete binder.

Many hempcrete binder formulations are mixtures of hydrated lime and hydraulic lime. It is also possible to replace the hydraulic lime with other hydraulic additives — known as *pozzolans* — such as metakaolin (fired kaolin clay), natural cement, blast furnace slag, fly ash, magnesium cement and Portland cement. The use of these pozzolans will be discussed in the chapter about mixes.

Sourcing of North American hydrated lime

Various brands of hydrated lime are available throughout North America from building supply stores or masonry supply outlets. There is some evidence that dolomitic lime is a better choice for hempcrete binders, but any hydrated lime that is locally available will work. Be sure to avoid agricultural lime (aglime), as this product has not been fired and is not reactive. Hempcrete binder made with agricultural lime will not work.

Cost of hydrated lime

Price can vary depending on distributor or retailer, and by quantity. Currently advertised prices range from $10–15 for a 50-pound (22.5 kg) bag. While ratios for hydrated lime to pozzolan can vary, for a mix that uses 50% hydrated lime the cost of the hydrated lime portion of the binder is $1.25 to $1.88 per cubic foot of wall insulation.

Sourcing and cost of pozzolans

There are numerous substances that have a pozzolanic reaction with hydrated lime:

Natural Hydraulic Lime (NHL) It is possible to create a custom mix of hydrated lime and NHL as a hempcrete binder. European NHL is available from specialty masonry supply outlets. Price can vary widely depending on sources, with retail listings ranging from $24–46 for a 55-pound bag.

Metakaolin or Calcined Kaolin Kaolin clay that has been fired at a high temperature (1200 to 1750°F or 650 to 900°C) can have a pozzolanic reaction with lime. It is often marketed for its pozzolanic qualities with cement and/or lime (called High Reactivity Metakaolin — HRM), and any product that conforms to ASTM C618, Class N pozzolan specifications should work in a hempcrete binder. Sourcing can be difficult, as it may not be available through typical masonry outlets. It is more typically used on a large, industrial scale and is often sold in bulk rather than individual bags. Prices can range widely, from $0.10 to $1.00 per pound.

Portland Cement I am frequently asked why hempcrete isn't just made using hemp hurds and Portland cement, and it's a good question. Portland cement is widely available, has a fast hydraulic set and creates a strong bond. From a performance point of view, mixes made with Portland cement end up being much denser than those made with lime and this lowers the thermal performance of the composite. Cement also has much poorer moisture-handling qualities than lime, making the material less permeable and unable to react as well to changes in humidity. It also lacks the degree of antimicrobial and antifungal properties of lime. Portland cement also carries significant embodied energy and carbon footprint implications. If Portland cement is used, it should be no more than 25% of the binder content. Retail costs of $5.80–8.00 for a 50-pound (22.5 kg) bag are typical.

Magnesium Cement The use of magnesium cement as a binder for hempcrete is at an experimental stage, but shows promise.[3] Strength and density using MgO binder are similar to lime-based binders, but the MgO has lower environmental impacts, including a lower carbon footprint. Currently, costs are higher and availability is more limited compared to other binders, but this material represents an interesting development for the future of hempcrete.

Proprietary Binders

There are two brands of formulated lime binder for making hempcrete:

Tradical Hemcrete® This product is from the UK. The safety data sheet for the binder shows that it is composed of "Calcium Hydroxide and Hydraulic Binder. Lesser quantities of calcium carbonate, calcium silicates, silica and oxides of magnesium, aluminium and iron and other trace elements." The binder has been formulated specifically for hempcrete, and includes hydrated lime (calcium hydroxide) as the lead ingredient. The "hydraulic binder" is not specified, so it could be any pozzolan or blend of pozzolans.

Batichanvre This product is from France, and comes from the world's largest producer of natural hydraulic lime. The safety data sheet for the binder lists calcium oxide (30–60% by weight), hydraulic binder (20%), calcium carbonate (10–30%), crystalline silica (5–10%), calcium hydroxide (~3%) and aluminum oxide (0.5–1.5%). The "hydraulic binder" in this case is likely to be the company's hydraulic lime.

Sourcing of proprietary binders

Specialty masonry distributors import both products in bagged form from Europe, with availability in both Canada and both coasts of the U.S.

Cost of proprietary binders

Price can vary depending on distributor or retailer, and by quantity. Currently advertised prices range from \$26 to \$38 per 55 lb (25 kg) bag. At the ratios suggested for wall mixes (1 bag of binder for ~4 ft^3 of hemp hurd) the cost of proprietary binders is \$6.50 to \$9.50 per cubic foot of wall insulation. The ratio of binder to hemp is lower for attic insulation and higher for slab insulation, with a resulting impact on price per cubic foot.

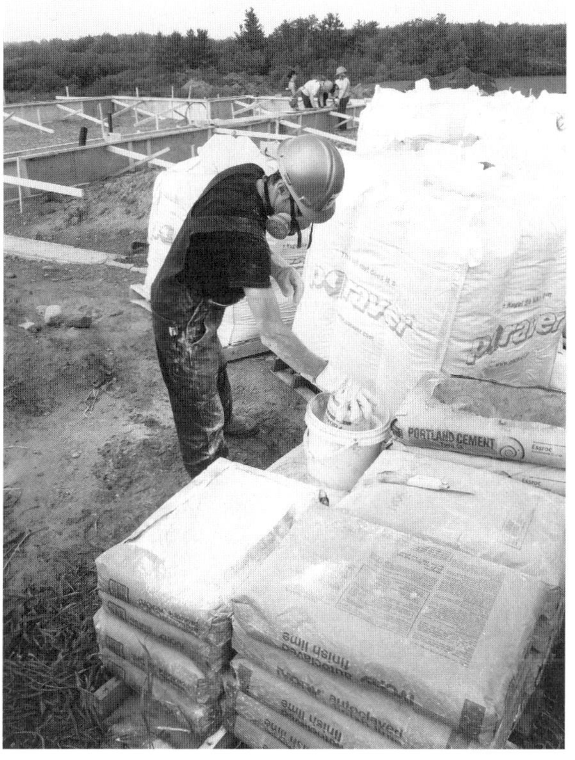

Metakaolin (Metapor from Poraver, in this case) and other pozzolans often come in Super Sacs rather than individual bags. Cost and ease of use are often better in this larger format.

Table 5.2: Environmental impacts of lime

Ecosystem Impacts	Embodied Energy	Carbon Footprint	Indoor Environment	Waste
Moderate. Limestone is a non-renewable resource but is abundantly available. Large scale quarrying can cause habitat destruction and surface and ground water interference and contamination.	**High.** 5.3 MJ/kg. Lime and/or cement are processed at high temperature, in addition to quarrying and crushing energy input.	**High.** 0.78 kgCO$_2$e/kg. Lime will absorb CO_2 during the curing process, but due to fuel use during processing will still be a net carbon emitter, though accurate figures are difficult to assess.	**Very low.** Lime-based plasters can contribute to high indoor air quality, providing naturally antiseptic qualities and no toxic off gassing.	**Low to Moderate.** Construction: Lime can be stored or used as a soil amendment. End of life: Lime-hemp can be broken up and recycled into a new mixture of hempcrete.

Wood Framing

A wood frame of some sort is a key element in hempcrete wall systems. Variations on the style of wood framing are covered in the Design Options chapter. Wood framing is also the most common option for roof framing if hempcrete is being used as roof insulation.

Strength and dimensions

Wood framing members for hempcrete walls and roofs should conform to the standards of IRC R602 (US) or NBC 9.3.2 (Canada). For use of wood in a structural role that does not conform to code prescriptions, *The Wood Handbook: Wood as an Engineering Material* from the Forest Products Laboratory of the USDA (FPL-GTE-190), *The Engineering Guide for Wood Frame Construction* from the Canadian Wood Council, or similar guidelines should be used.

Sourcing

Wood framing materials can be sourced from conventional building supply outlets or directly from local sawmills. Wood suppliers with third-party sustainability certifications (such as FSC — Forest Stewardship Council) are likely to have lesser ecosystem impacts.

Cost

Wood framing retail costs range from \$0.32–0.51 per lineal foot (\$1.05–1.67 per m) for 2×4s to \$0.50–0.63 per lineal foot (\$1.64–2.07 per m) for 2×6s.

The use and arrangement of lumber in hempcrete walls will vary according to loads, code requirements and panel type, and costing will need to reflect a specified design.

Table 5.3: Environmental impacts of wood framing

Ecosystem Impacts	Embodied Energy	Carbon Footprint	Indoor Environment	Waste
Low to High. Forestry practices can range from third-party verified sustainable harvesting to unregulated clear cutting. Confirm practices with source to verify degree of impact.	**Low.** 7.4 MJ/kg* (spruce lumber) or 43.66 MJ per 2x4x8. Quantities of lumber used for different prefabricated wall systems will vary widely, and total embodied energy figures must be assessed based on design.	**Low.** 0.59 $kgCO_2e/kg$.* Quantities of lumber used for different hempcrete wall systems will vary widely, and total carbon footprint must be assessed based on design. High carbon sequestration potential.	**Low.** Framing lumber in most panel systems is not in direct contact with indoor air, but softwood lumber does not have toxic off gassing or contain any red list chemicals.	**Low to High.** Construction: Framing lumber can be utilized strategically to minimize waste, but standard lengths can lead to high percentage of off cuts. Wood waste can be recycled or composted. End of life: Can be recycled or composted. Will require separation from assembly.

Note:
* Data is from Inventory of Carbon and Energy (ICE) 2.0, University of Bath

Sheathing Materials

Lime plaster

Lime plaster (or *render,* as it's known in Europe) is the most common sheathing for interior and exterior faces of hempcrete walls and includes formulas that use either hydraulic or hydrated lime as the binder. In general, these plasters use a ratio of binder to well-graded aggregate in the range of 1 part binder to 2–4 parts aggregate. Fiber is a common additive in these mixes, and can include poly fibers, chopped straw or animal hair.

Thickness

Lime plaster should have a maximum thickness of ½ inch (16 mm) per coat.

Strength

Lime plaster should have a compressive strength of 600 psi (4.14 MPa) at 28 days as per IRC Appendix S, AS106.12.

Permeability

1:3 lime:sand — 9 US perms (500 metric perms)[4]

Basic lime plaster formulation (by volume):

1 part hydrated lime or hydraulic lime (NHL 3.5 or 5)

2–3 parts sharp aggregate

0–1 part fiber

Water to provide a plastic, workable mix

Sourcing

See lime for binder.

Cost

Hydrated lime prices range from $10–15 for a 50-pound (22.5 kg) bag. At a thickness of 1 inch on both sides of a panel, the cost will be $0.69–1.03 per square foot of wall ($7.42–11.08 per m^2). Sand will need to be added to hydrated lime to make a plaster, at an average cost of $0.09–0.12 per square foot ($0.97–1.29 per m^2).

Table 5.4: Environmental impacts of lime plaster

Ecosystem Impacts	Embodied Energy	Carbon Footprint	Indoor Environment	Waste
Moderate. Limestone is a non-renewable resource but is abundantly available. Large scale quarrying can cause habitat destruction and surface and ground water interference and contamination.	**High.** 1.11 MJ/kg* or 116.9 MJ/m^2. Lime is processed at high temperature, in addition to quarrying and crushing energy input.	**High.** 0.174 kgCO_2e/kg* or 18.33 kgCO_2e/m^2. Lime will absorb CO2 during the curing process, but due to fuel use during processing will still be a net carbon emitter, though accurate figures are difficult to assess.	**Very low.** Lime-based plasters can contribute to high indoor air quality, providing naturally antiseptic qualities and no toxic off gassing.	**Low to Moderate.** Construction: Plasters can be left in the environment or crushed to make aggregate. End of life: Plasters can be left in the environment or crushed to make aggregate. Embedded mesh will require separation.

Note:
* Data is from Inventory of Carbon and Energy (ICE) 2.0, University of Bath

Wood fiber sheathing

Wood fiberboard sheet materials can be used as insulative and structural exterior and/or interior sheathing for hempcrete walls, and can be placed as permanent formwork (or shuttering) on one side of the wall.

This sheathing material is made of compressed wood fibers (typically recycled) compressed, with a wax-based binder. Wood fiberboard products used for hempcrete walls should conform to the standards of ASTM E72 (Standard Test Methods of Conducting Strength Tests of Panels for Building Construction) and/or ASTM C209 (Standard Test Methods for Cellulosic Fiber Insulating Board) or equivalent. Because they are permeable and can provide structural and insulative properties, they are a good option.

Many fiberboard products are rated for exterior applications, but some are not. Be sure to specify exterior grades if that is what you require.

Strength

The strength requirements for wood fiberboard sheathing will vary depending on the design of the hempcrete wall. The sheathing may be designed to play a large role in the structural performance of the wall, or it may not figure into the structural calculations at all. The ASTM E72 and/or C209 data can be used to ensure that a particular product will meet the strength requirements.

Should a frame not have anchoring points that match the requirements of established performance standards, engineering principles should be used to determine appropriate fastening methods and loads.

Dimensions

North American products typically conform to standard sheet sizes (4×8 and 4×9 feet). European imports are often nonconforming, and should be confirmed prior to ordering

Table 5.5: Environmental impacts of wood fiber sheathing

Ecosystem Impacts	Embodied Energy	Carbon Footprint	Indoor Environment	Waste
Low to Moderate. Most wood fiber products are made from post-industrial waste streams and do not directly involve the harvesting of timber. Third-party verified practices can be sourced, and practices should be confirmed with source to verify impacts. Most manufacturers use nontoxic binders; this should be confirmed.	**Moderate.** 9.36MJ/kg** or 71.3 MJ/m^2 at 1-inch. There are both wet and dry processing methods for wood fiber sheathing, and no third-party data is currently available to assess this category thoroughly. The additional energy efficiency added by these types of panels helps off set EE compared to plaster or non-insulating sheathing.	**Low.** There are no available figures for this product category. Particle board is rated at 0.86 $kgCO_2e/kg$* and uses similar processing techniques. Carbon sequestration potential for this material is high.	**Low to Moderate.** These products are made using a variety of binder materials. Many manufacturers advertise nontoxic binders, and some have third-party verification for low emissions. Sheathing is not in direct contact with indoor air.	**Low to High.** Construction: Sheathing can be utilized strategically to minimize waste, but standard sizes can lead to high percentage of off cuts. Wood waste can be recycled or composted. End of life: Can be recycled or composted. Will require separation from assembly.

Note:
* Data is from Inventory of Carbon and Energy (ICE) 2.0, University of Bath
** From Environmental Building News, Assessing Sheathing Options

Panels can come in a wide range of thicknesses, from ½ inch to 4 inches (12.7 to 101.6 mm). Square edges or tongue-and-groove edges may be available.

Thermal performance

Wood fiberboard products typically have thermal resistance values of R-2.5 to 3.5 per inch. Check with manufacturer for product-specific ratings.

Permeability

Thickness and density of products will affect perm ratings. Permeance in the range of 18–28 US perms is common, qualifying them as fully vapor permeable. Wood fiberboard products can be used to create a permeable sheathing on either or both sides of a hempcrete wall.

Sourcing

Wood fiberboard products are currently well developed in European markets, and less so in North America.

The North American Fiberboard Association (fiberboard.org) represents the industry, and can be used to source local manufacturers and distributors of products.

European companies have limited distribution in North America, typically through specialized green building product outlets. Fiberboard is a common material used in Passive House projects in Europe, and distributors may be located via local Passive House chapters.

Cost

Price for wood fiberboard will vary by thickness. Products with structural properties at 1½ inch (38 mm) thickness range from $1.00 to $2.00 per square foot ($10.76 to 21.52 per m^2).

Magnesium oxide sheathing

Magnesium oxide (MgO) sheathing is a relative newcomer to the construction industry. This sheet material is made from magnesium oxide cement cast into thin cement panels and cured under proper conditions. Most products use a percentage of wood chips and/or perlite in the mix, and the boards are commonly faced with fiberglass matts on both sides. MgO board has a relatively high structural capacity, excellent fire rating, and is resistant to mold and mildew.

Magnesium oxide sheathing products can be used as an exterior and/or interior sheathing for hempcrete walls, and can be placed as permanent formwork on one side of the wall if the permeability is 7 US perms or higher.

Strength

The strength requirements for MgO sheathing will vary depending on the design of the wall system. There are not yet specific ASTM standards for MgO board, but structural capacity can be found via results for ASTM E72 (Standard Test Methods of Conducting Strength Tests of Panels for Building Construction).

Many MgO products have not undergone ASTM or equivalent testing for use as a structural sheathing. While these products may be strong enough, it is best to choose products that have passed structural tests if they are to be used for structural purposes.

Should a hempcrete wall not have anchoring points that match the requirements of established performance standards, engineering principles should be used to determine appropriate fastening methods and loads.

Dimensions

MgO panels come in a variety of standard sizes. Those intended to be used as structural sheathing are commonly 4×8, 4×9 or 4×10 feet (1220×2400, 1220×2750 or 1220×3050 mm) at ½ and ⅝ inch (12.7 and 16 mm) thicknesses. MgO products used as tile backer are commonly 3×5 feet (915×1525 mm).

Table 5.6: Environmental impacts of magnesium oxide sheathing

Ecosystem Impacts	Embodied Energy	Carbon Footprint	Indoor Environment	Waste
Moderate to High. Magnesium carbonate is quarried from surface-based pits, mostly in Asia. It is difficult to obtain accurate information about impacts. Large scale quarrying can cause habitat destruction and surface and ground water interference and contamination.	**Moderate to High.** 6 MJ/kg* or 56.5 MJ/m² at ½". There are no third-party figures available. MgO is heated during processing, resulting in relatively high EE. Fiberglass mesh is integrated, as is perlite. All of these high-intensity materials make the industry figure quoted seem too low.	**Moderate to High.** There are no available figures for this product category.	**Low.** This product category uses stable, nontoxic basic materials and reports to be free of off gassing and toxic chemicals. There are currently no third-party verifications, but MgO products are recommended by certified Bau-Biologists.	**High.** Construction: Sheathing can be utilized strategically to minimize waste, but standard sizes can lead to high percentage of off cuts. Composite material cannot be composted or recycled. End of life: Cannot be recycled or composted. Will require separation from assembly.

Note:
* Data is from Inventory of Carbon and Energy (ICE) 2.0, University of Bath

Thermal performance

MgO panels do not contribute meaningfully to the thermal performance of a hempcrete wall. No manufacturers rate the thermal performance of the board at this time. However, the product can contribute to thermal performance if it is used to provide an effective air control layer for the interior and/or exterior of the panel.

Permeability

MgO sheathing used on the interior face of hempcrete walls should be higher than 2 US perms, and the exterior face should rate higher than 4 US perms. Permeability values for MgO board from different manufacturers range widely, from as low as 0.9 to 7.5 US perms. Given the wide variation, be sure to confirm perm ratings with manufacturers prior to ordering material. MgO used as permanent forms should be at least 7 US perms.

Sourcing

Most of the world's magnesium board comes from China, though some North American production is beginning to get underway. Magnesium board is available through some conventional building supply outlets and distributors, though often as a special order and not a stock item. It may be beneficial to go directly to manufacturers/importers.

Quality control has been an issue for many magnesium board suppliers. Due diligence should be performed to ensure that consistent quality is available from the chosen supplier.

Cost

MgO board at ½ inch (12.7 mm) thickness range from $0.80 to $1.20 per square foot, retail price.

Gypsum sheathing

Gypsum sheathing comes in two forms, one for exterior use (featuring a fiberglass and/or waxed paper coating and a wax-treated gypsum core) and one for interior use (conventional "drywall"). Both are appropriate for use with hempcrete walls, but only the exterior product should be used as permanent formwork. Indoor gypsum board can be applied to a hempcrete wall as the sheathing material only after the hempcrete has fully dried.

Strength

Gypsum sheathing products used as an exterior and/or interior sheathing for hempcrete walls should conform to the standards of ASTM C1396 (Standard Specification for Gypsum Board), ASTM C1278 (Standard Specification for Fiber-Reinforced Gypsum Panel), or ASTM C1177 (Standard Specification for Glass Mat Gypsum Substrate for Use as Sheathing).

Should a hempcrete wall not have anchoring points that match the requirements of established performance standards, engineering principles should be used to determine appropriate fastening methods and loads.

Dimensions

Gypsum sheathing products come in a variety of standard sizes. Those intended to be used as structural sheathing are commonly 4×8, 4×9 or 4×10 feet (1220×2400, 1220×2750 or 1220×3050 mm) at ½ or ⅝ inch (12.7 or 16 mm) thicknesses. Thinner products may be used as an interior finishing layer on hempcrete walls.

Thermal performance

Gypsum sheathing products do not contribute meaningfully to the thermal performance of a hempcrete wall. The thermal performance of ½-inch (12.7 mm) board is typically R-0.45. However, the product can contribute to thermal

Table 5.7: Environmental impacts of gypsum sheathing

Ecosystem Impacts	Embodied Energy	Carbon Footprint	Indoor Environment	Waste
Moderate. Gypsum is a soft rock quarried from surface-based pits. Large scale quarrying can cause habitat destruction and surface and ground water interference and contamination.	**Moderate to High.** 6.75 MJ/kg* or 60.75 MJ/m² at ½ inch. Gypsum is processed using a moderate amount of heat. Fiberglass facing is applied, and may not be included in the above figure.	**Moderate.** 0.39 $kgCO_2e/kg$* or 3.51 $kgCO_2e/m^2$. Does not include production of fiberglass facing.	**Moderate.** *Exterior product:* Fiberglass particulate is shed during handling. The material may contain some quantity of toxic chemicals, including vinyl acetate monomer, acetaldehyde and formaldehyde. Product would be used only on exterior of wall. *Interior product:* The paper and glue of interior drywall can be a good medium for mold growth in wet conditions.	**High.** Construction: Sheathing can be utilized strategically to minimize waste, but standard sizes can lead to high percentage of off cuts. Composite material cannot be composted or recycled. End of life: Cannot be recycled or composted. Will require separation from assembly.

Note:
* Data is from Inventory of Carbon and Energy (ICE) 2.0, University of Bath.

performance if it is used to provide an effective air control layer for the interior and/or exterior of the panel.

Permeability

Interior and exterior gypsum products have tested permeance ratings of 20 to 25 US perms, making them suitable for use as a permeable sheathing.

Sourcing

Gypsum panels are widely available through conventional building supply outlets.

Notes

1. "Complete chemical analysis of Carmagnola hemp hurds and structural features of its components," Gandolfi et al. (2013). "Analysis of hemp hurds," *BioResources* 8(2), 2641–2656.
2. Based on bulk pricing from ten different online suppliers.
3. Lucia Kidalova et al. "MgO cement as suitable conventional binder replacement in hemp concrete," Institute of Building and Environmental Engineering, Technical University of Kosice, Kosice, Slovakia, 2011.
4. *Design of Straw Bale Buildings,* Bruce King et al., Green Building Press, pg 156.

Chapter 6

Design Options for Hempcrete

THIS CHAPTER OUTLINES A VARIETY OF OPTIONS for hempcrete walls that a designer and/or builder can consider before moving forward with a project. These options are based on the work of hempcrete builders around the world; the final details used for any of these types of construction will need to be aligned with local code requirements and specific project needs.

There is no single option that is better than any other. By clearly identifying the projects needs in terms of material costs and availability, labor input, structural and thermal requirements, there will be an option that will work for any project.

It must be remembered that a hempcrete wall is a combination of elements that together provide the complete structural, thermal, air, moisture and water protection requirements demanded of a wall. All elements must work in theory and in practice for the system to be successful.

To Plaster or Not?

One of the major questions to answer is whether or not to use a plaster finish on the interior or exterior of the wall system. Plaster can provide a permeable and durable finish, and hempcrete provides an excellent substrate for applying plaster. In Europe and the UK, the vast majority of hempcrete walls are plastered.

In North America, plastering can have a number of drawbacks:

- Plastering is not a common trade and therefore not a common finish in many regions. Where plaster — or "stucco" as it's typically called — is used, it tends to be an acrylic-based product that is not suitable for application on hempcrete walls due to its very low permeability. Material availability and knowledgeable installers of traditional permeable plasters may be difficult to source.
- Although material costs for plaster are low compared to other sheathing options, labor input can be much higher. Hempcrete walls provide a good substrate for plaster, but additional preparation (meshing or sand priming) is required where plaster will be applied over wood and other wall elements. Flashing and trim details can be involved, and while plaster can definitely provide a complete and effective air control layer for the wall, the details required to achieve high performance in this regard also add time.
- Plastering, especially on exterior surfaces, must be performed when the weather is above 40°F (5°C) and ideally less than 90°F (32°C), which puts limitations on construction schedules in many parts of the continent.
- Exterior plaster finishes can be prone to cracking. When cracks are not excessively wide (less than 3/64 to 5/64 inch, or 1 to 2 mm), they are not problematic, but the visual appearance of cracks seems to offend the North American sensibility. Plastering also adds an additional "wet process" to a wall system that already faces long drying times.
- Plastering requires the installation and removal of formwork, adding an additional step to the construction process.

None of these are reasons to avoid plaster finishes, especially on the interior side of a hempcrete wall. Exterior plastering is entirely viable, as long as the designer/builder is aware of the impacts the choice carries with it.

Permanent, Permeable Sheathing

An attractive option in the North American context is the use of permanent, permeable exterior sheathing. This material creates one side of the formwork when it is permanently attached to the frame during wall construction (as is common framing practice), and it allows for the installation of siding options that are more typical and simpler to source and install. The sheathing provides immediate rain protection for the walls and the building if installed on the exterior side.

The major drawback to permanent permeable sheathing is that it will slow down drying times to some extent, robbing one side of the wall of free air circulation. It also necessitates the use of an additional layer of material to provide the final weather protection layer for the building in the form of a rain screen siding.

A builder or designer should be aware of all the implications of choosing plaster or permeable sheathing for a hempcrete wall and make choices from the options outlined below based on the needs of the project.

Single Stud Framing

This approach uses code-compliant framing techniques to create a structural wall. The use of a code-compliant framing system is a definite advantage when it comes to permitting, and it means that conventional framing crews can be hired to create the structure for a hempcrete wall. There are several ways that this type of wall can be designed, each of which offers different thermal performance.

Simple single stud framing

A frame is built of 2×4, 2×6 or 2×8 lumber and the stud cavities are filled with hempcrete insulation. The stud spacing for the wall can conform to local codes, or can be determined by a structural engineer to take advantage of the structural support offered by the hempcrete. The use of a permeable, insulated sheathing for this style of wall is recommended to minimize thermal bridging and the weakening of thermal performance this can cause.

 ***Advantages:***

- Simplicity of frame construction, matches code and trades expectations.
- Simplicity of attaching formwork/shuttering; wood framing is exposed on both sides of insulation.
- Multiple layers of insulated sheathing may be added to increase thermal performance.
- Ventilated rain screen siding easily accommodated.
- Narrow wall profile matches conventional expectations.

 ***Disadvantages:***

- Thermal performance limited to depth of insulation achieved within the standard framing dimensions.
- Thermal bridging will reduce thermal performance, although the use of insulated sheathing will help to mitigate.

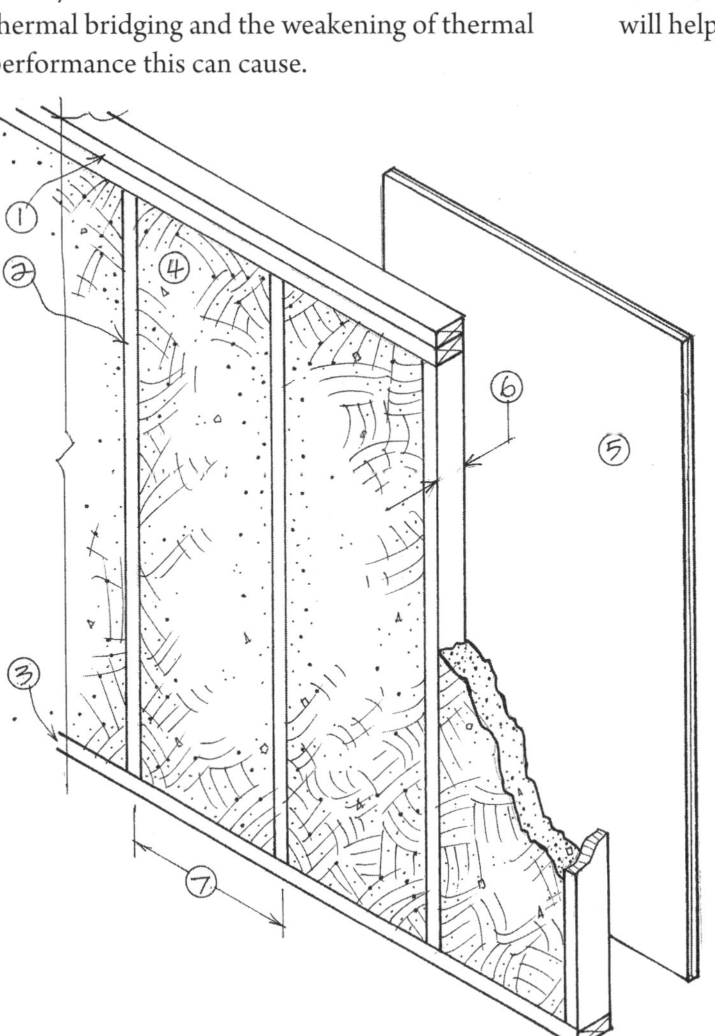

1. Doubled top plate
2. Stud (2×4, 2×6 or 2×8)
3. Sill plate
4. Hempcrete insulation @ 300kg/m^3
    - 3½ inches @ R-2/inch = R-7 (plus sheathing R-value)
    - 5½ inches @ R-2/inch = R-11 (plus sheathing R-value)
    - 7¼ inches @ R-2/inch = R-14.5 (plus sheathing R-value)
5. Plaster adhered to hempcrete, or structural sheathing (ideally insulated sheathing) fastened to framing. Insulated sheathing can be used on both sides of the frame to increase thermal performance. Multiple layers of insulated sheathing may be used to further increase thermal performance.
6. Wall framing width determined by codes and desired thermal performance
7. Stud spacing determined by codes or engineering optimization

Single stud framing with furring

A simple stud wall can have horizontal furring attached to it to provide a wider cavity for insulation, without needing to build a second wall.

 ***Advantages:***

- Simplicity of frame construction, matches code and trades expectations.
- Additional furring a straightforward addition.
- Furring adds insulation depth and greatly reduces thermal bridging.
- Simplicity of attaching formwork/shuttering; wood framing is exposed on both sides of insulation.
- Multiple layers of insulated sheathing may be added to increase thermal performance.
- Ventilated rain screen siding easily accommodated.
- Electrical wiring can be routed along furring without need for drilling holes in studs.

 ***Disadvantages:***

- Thermal bridging at points of contact between studs and furring, although the use of insulated sheathing will help to mitigate.
- Thermal performance limited to depth of insulation achieved within the standard framing dimensions (assuming 2×3 furring).
- Additional furring at window and door openings can add complication.

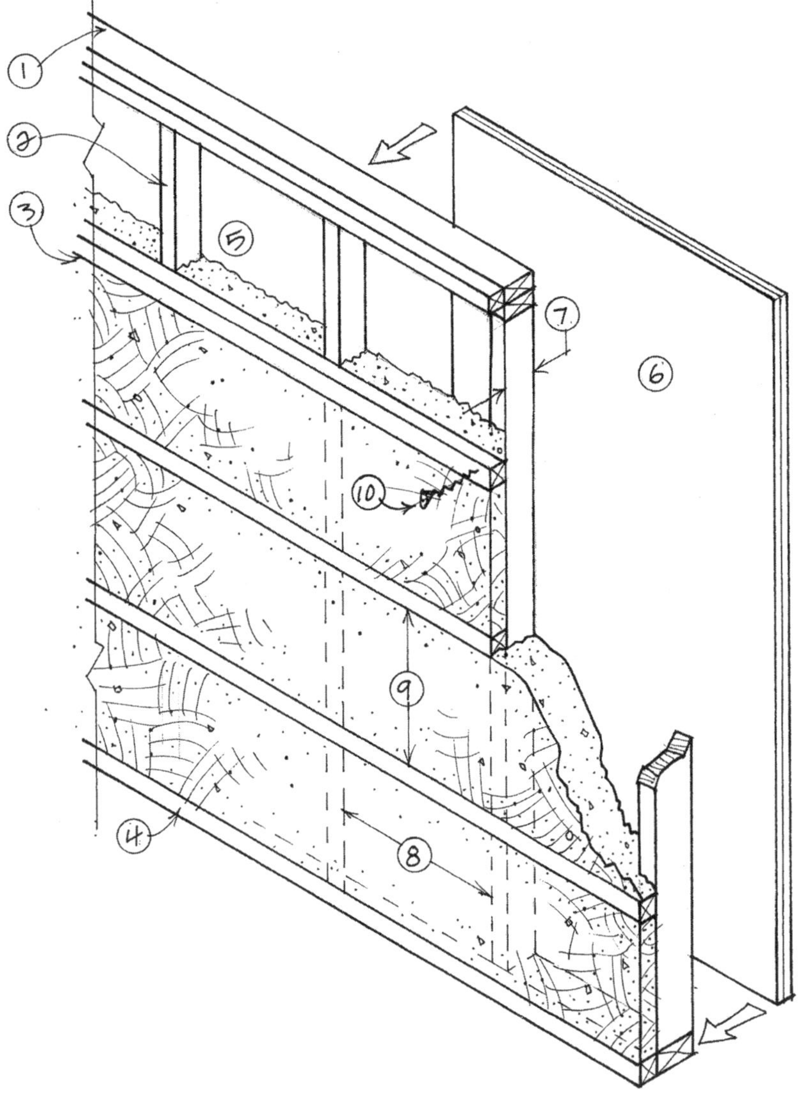

1. Doubled top plate
2. Stud
3. Horizontal furring (2x2 or 2x3)
4. Sill plate
5. Hempcrete insulation @ $300 kg/m^3$
   - 3½ + 2½ inches @ R-2/inch = R-12 (plus sheathing R-value)
   - 5½ + 2½ inches @ R-2/inch = R-16 (plus sheathing R-value)
   - 7¼ + 2¼ inches @ R-2/inch = R-19 (plus sheathing R-value)
6. Plaster adhered to hempcrete, or structural sheathing (ideally insulated sheathing). Insulated sheathing can be used on both sides of the frame to increase thermal performance. Multiple layers of insulated sheathing may be used to further increase thermal performance.
7. Wall framing width determined by codes and desired thermal performance
8. Stud spacing determined by codes or engineering optimization
9. Furring spacing at 16-inch or 24-inch, depending on sheathing specifications
10. Fastener appropriate to width of furring material

Single stud framing centered in hempcrete

This approach matches the simplicity of single stud framing with the potential for creating any thickness of hempcrete insulation. This system is used quite frequently in the UK, and plaster is the most sensible finish for the walls.

 ***Advantages:***

- Thermal performance can be determined independently of framing size.
- No thermal bridges.
- Plaster finishes can be applied to an uninterrupted hempcrete surface, with no need to mesh exposed wood.

 ***Disadvantages:***

- Formwork/shuttering is more complicated and time consuming, as forms cannot attach directly to framing.
- Placement of windows and doors must conform to placement of frame, or additional framing must be built to accommodate alternative placement.
- Windowsills, returns and trim more difficult to achieve.
- More difficult to run electrical wiring and fasten electrical boxes.
- No backers to attach trim at the bottom or top of the wall unless additional framing is added.
- Plaster is the only finishing option, as there is no easy way to attach sheathing and/or siding.

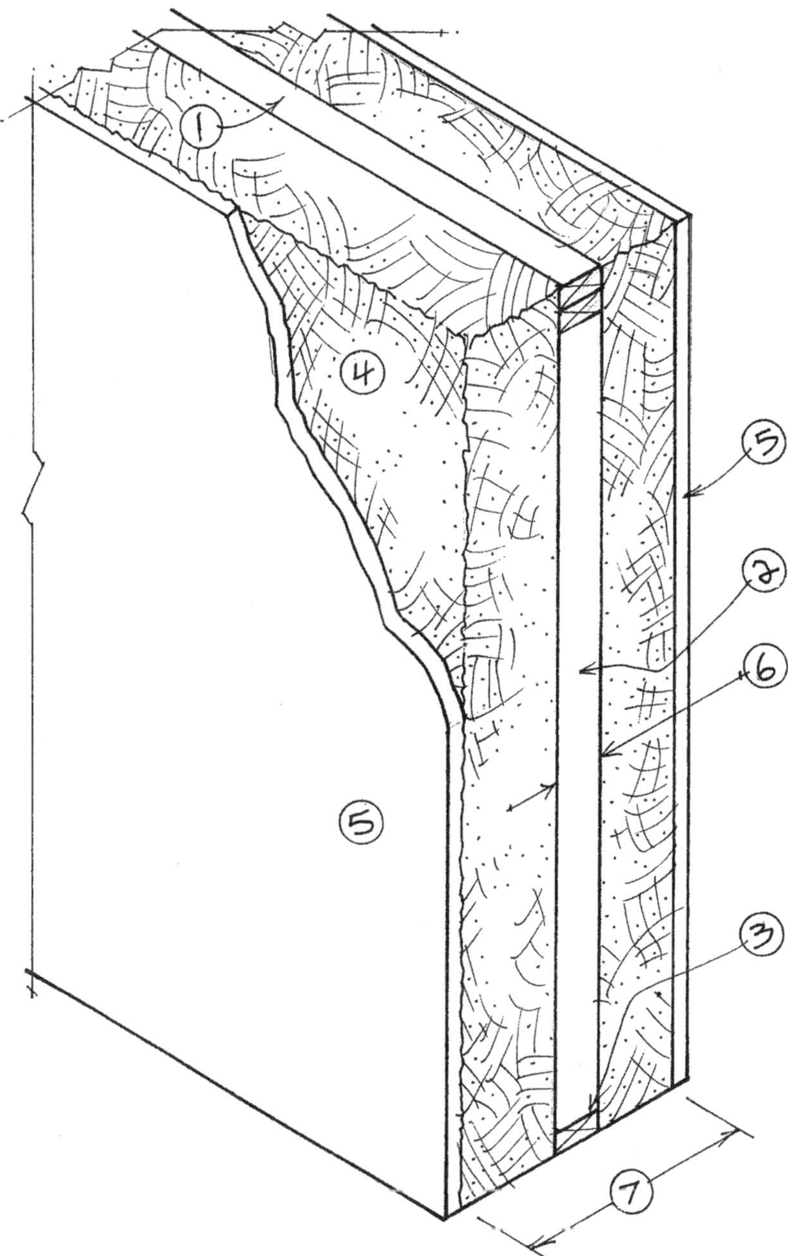

1. Doubled top plate
2. Stud (spacing determined by code or engineering optimization)
3. Sill plate
4. Hempcrete @ 300kg/m^3 encapsulating frame (frame could be centered as shown, or be flush to the exterior or the interior)
5. Plaster applied directly to hempcrete surface (if frame is flush on one side, sheathing material can be fastened to that side)
6. Width of framing does not affect thickness of insulation, so can be optimized for structural purposes only
7. Width of hempcrete can be optimized to suit thermal performance independent of frame size

Double Stud Framing

This approach uses code-compliant framing techniques to create a structural wall, and a second (usually nonstructural, but can be used to share loads) wall to create the desired depth of insulation.

Double stud framing with full hempcrete cavity

This is the most typical version of double stud framing, with the frame providing attachment for formwork and/or sheathing on both sides.

 ***Advantages:***

- Formwork/shuttering easy to attach to frame.
- Thermal performance can be determined independently of framing size.
- Framing can be sized so exterior wall handles roof loads and interior wall handles floor joist loads.
- No thermal bridges.
- Ventilated rain screen siding easily accommodated.
- Easy to run electrical wiring and fasten electrical boxes on both sides of the wall.
- Easy to fasten trim and frame for window sills and returns.
- Both sides can be plastered or sheathing can be fastened to framing.
- Wood use can be similar to single stud, despite two frames.
  - Single 2×6 wall has 5½ inches of wood depth.
  - 2×4 wall plus 2×3 wall has 6 inches of total wood depth.

 ***Disadvantages:***

- Additional framing time and cost.

1. Double top plate
2. Stud (spacing determined by code or engineering optimization)
3. Sill plate
4. Secondary top plate (double if bearing structural loads, single if non-load bearing)
5. Secondary stud (spacing to match structural wall, or optimized for other purposes)
6. Secondary sill plate
7. Hempcrete insulation @ 300 kg/m^3
8. Plaster adhered to hempcrete or sheathing fastened to framing. Both sides can match, or can use different option on each side.

Double stud framing with interior service cavity

A double stud frame is built with a permanent sheathing on the exterior side of the interior frame, keeping the hempcrete formed to the back side of the interior wall. This leaves the interior frame wall as an open cavity for running wiring and plumbing without being buried in hempcrete.

 ***Advantages:***

- Formwork/shuttering easy to attach to frame.
- Thermal performance can be determined independently of framing size.
- Framing can be sized so exterior wall handles roof loads and interior wall handles floor joist loads.
- No thermal bridges.
- Ventilated rain screen siding easily accommodated.
- Mechanical services are not buried in hempcrete and do not penetrate interior air barrier. Future alterations do not require removal of hempcrete.
- Easy to fasten trim and frame for window sills and returns.
- Exterior can be plastered or sheathing can be fastened to framing.
- Wood use can be similar to single stud, despite two frames.
  - Single 2×6 wall has 5½ inches of wood depth.
  - 2×4 wall plus 2×3 wall has 6 inches of total wood depth.

 ***Disadvantages:***

- Additional framing time and cost.
- Additional layer of sheathing required to finish interior of wall.

1. Double top plate
2. Stud (spacing determined by code or engineering optimization)
3. Sill plate
4. Secondary top plate (double if bearing structural loads, single if non load bearing)
5. Secondary stud (spacing to match structural wall, or optimized for other purposes)
6. Secondary sill plate
7. Sheathing fastened to exterior side of interior wall frame
8. Hempcrete insulation @ 300 kg/m^3
9. Plaster adhered to hempcrete or sheathing fastened to framing.
10. Open cavity for services, to be sheathed with a final finish material

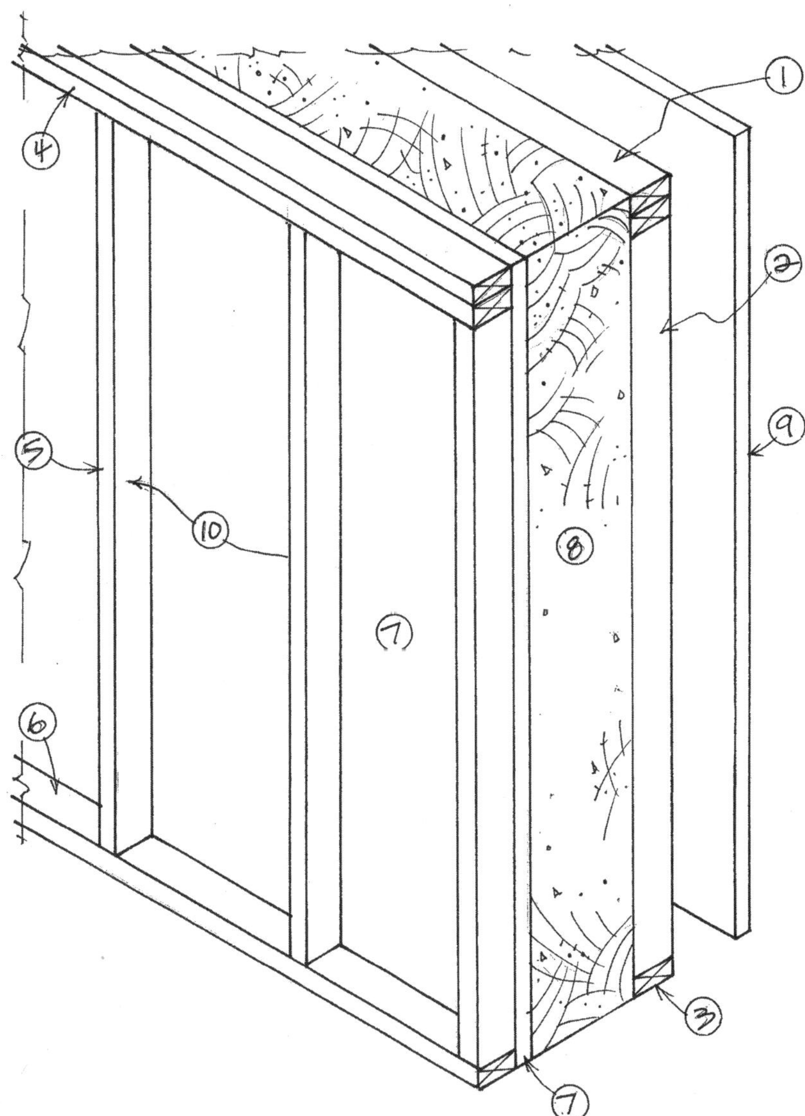

Timber Frame or Post and Beam

A structural frame can be constructed with heavy timbers, using traditional wooden joinery (timber framing) or metal connectors and fasteners (post and beam). The hempcrete can be formed within this frame, or can wrap the frame. In some cases, very light, nonstructural wooden framing is used to help create the hempcrete wall and facilitate formwork/shuttering.

Hempcrete within heavy timbers

The timber framework can be wrapped in hempcrete insulation, and typically the face of the timber is left to show to the interior of the building. If the timbers are not intended to be a part of the aesthetic of the building, they can be completely buried in the hempcrete (see Single Stud Framing Centered in Hempcrete, above, for details). The amount that the timbers are exposed can be determined according to aesthetic and/or thermal performance.

 ***Advantages:***

- No thermal bridges.
- Thermal performance can be determined independently of framing size.
- Few interruptions/junctions for easier hempcrete installation.
- Aesthetic possibilities of heavy timbers.
- Consolidation of building structure, insulation can be removed/replaced without affecting structural integrity.

 ***Disadvantages:***

- Window and door framing must be built independently of timber frame.
- Temporary or permanent framing may need to be created for formwork/shuttering.
- Additional framing time and cost for heavy timber work.
- Air sealing details around heavy timbers must be addressed to prevent leaks around timbers where plaster or sheathing intersect with wood to ensure airtightness and thermal performance.

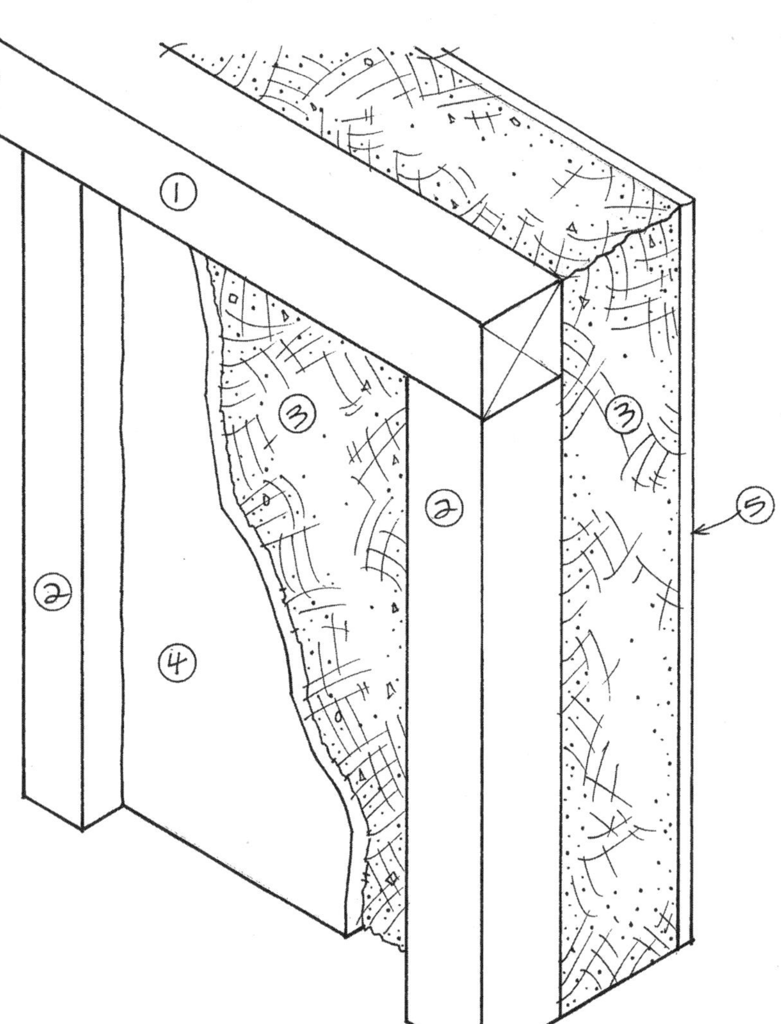

1. Beam
2. Posts
3. Hempcrete insulation @ 300kg/m3
4. Interior plaster or sheathing
5. Exterior plaster

Air Sealing Details for Timber Framing

A

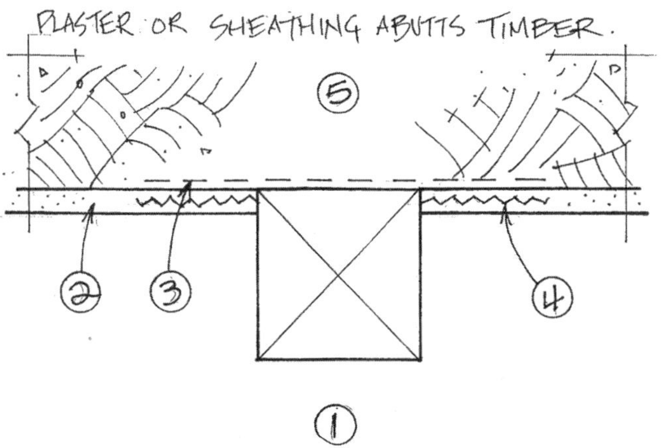

1) Timber
2) Plaster or sheathing
3) Sheet-style air barrier runs continuously behind timbers and extends onto hempcrete face
4) Mesh over air barrier for plaster only
5) Hempcrete

B

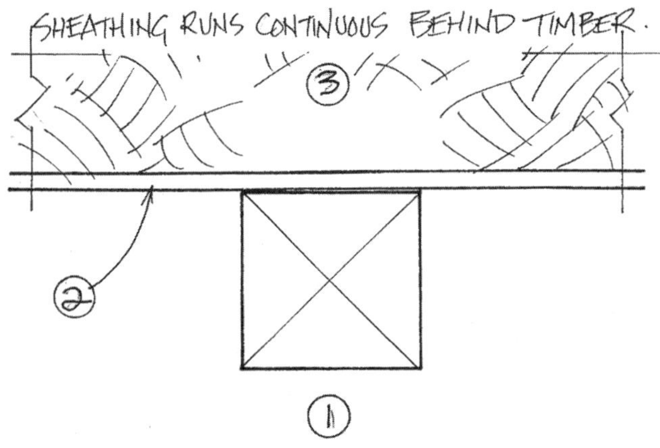

1) Timber
2) Sheathing runs continuously behind timbers, can be used as permanent formwork
3) Hempcrete

C

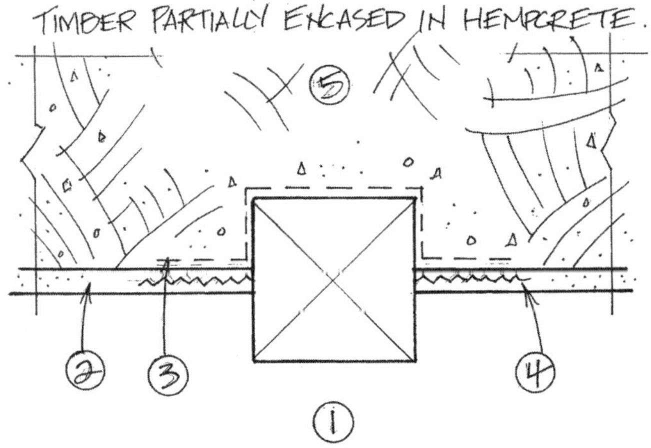

1) Timber
2) Plaster or sheathing
3) Sheet-style air barrier runs continuously behind timbers and extends onto hempcrete face
4) Mesh over air barrier for plaster only
5) Hempcrete

D

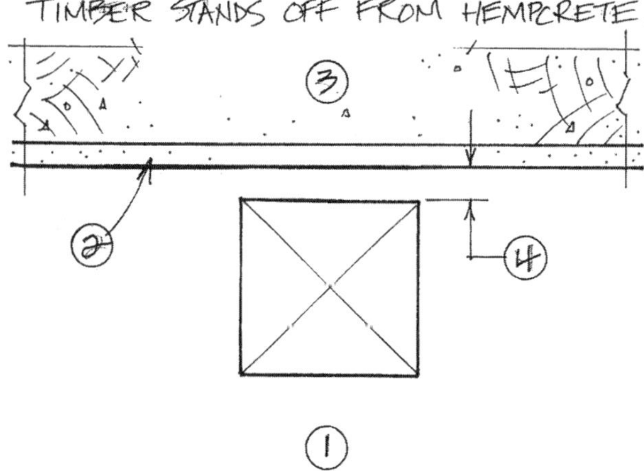

1) Timber
2) Plaster or sheathing uninterrupted by timbers
3) Hempcrete
4) Gap must be large enough to accommodate placement of sheathing or plastering

Design Options for Hempcrete Roof Insulation

Hempcrete insulation can be adapted to most styles of roof framing. The additional weight of hempcrete roof insulation (200 to 250 kg/m^3, or 12.5 to 15.6 lb/ft^3, compared with ~30 kg/m^3 or 1.9 lb/ft^3 for lightweight insulation materials) requires accommodation at the design stage, and will likely result in more robust roof framing. Due to the relatively low thermal resistance of the hempcrete (R-2 to 2.4 per inch, compared with R-3.5 to 5 per inch for lightweight options), roof insulation will need to be deep enough to meet thermal performance requirements, and this extra depth needs to be built into the roof framing design.

The options presented here are not the only possibilities, but they represent a range of commonly used styles. Other roof framing styles may be suitable, as long as they meet all structural and thermal performance requirements.

Manufactured Truss Roofs

Standard truss roof

In some cases, a typical truss roof design may be adaptable for use with hempcrete roof insulation. This type of truss provides a flat ceiling.

 ***Advantages:***

- Simple truss design, conforms to conventional design standards and builder expectations.
- Truss roof typically has lower material and installation costs than other roof framing options.
- Truss roofs can use smaller dimension lumber than other roof types.

 ***Disadvantages:***

- Thermal bridge at each truss bottom chord.
- Pitch may need to be steep to achieve appropriate insulation depth over wall.
- Custom insulation baffles must be installed between each truss; must be adequate to contain hempcrete.

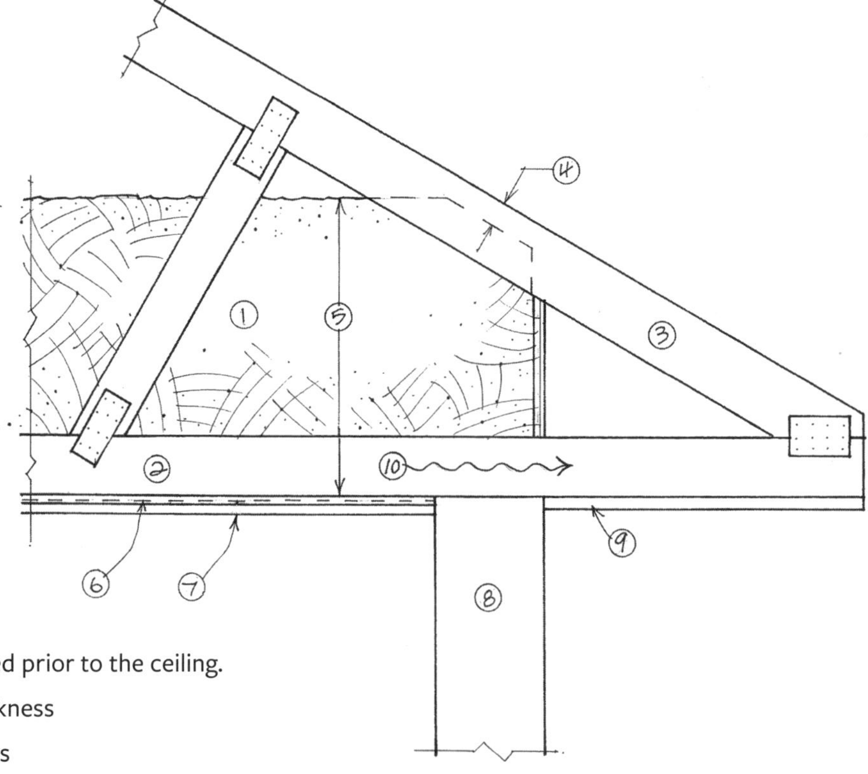

1. Hempcrete insulation @ 200 kg/m^3
2. Truss bottom chord, designed to accommodate dead load of hempcrete
3. Truss top chord, designed for appropriate loads
4. Insulation baffle with code-approved ventilation space over insulation
5. Full depth of insulation must be reached directly over the wall to meet thermal performance requirements. Adjust roof pitch to accommodate.
6. Vapor control layer below insulation to prevent excessive moisture accumulation in attic area, as per local code requirements. Control layer wraps onto wall to provide air sealing at junction.
7. Finished ceiling, able to handle dead load of hempcrete. Insulated sheathing could be applied prior to the ceiling.
8. Hempcrete wall of appropriate design and thickness
9. Ventilated soffit, as per local code requirements
10. Thermal bridge to exterior at each truss bottom chord

Raised heel truss

A common truss design for high-performance homes, this type of truss solves several issues for designs with thick roof insulation. This type of truss provides a flat ceiling.

 ***Advantages:***

- Full depth of insulation over the wall can be achieved with any roof pitch.
- Minimized thermal bridging from roof framing.
- Simple insulation baffling created by insulat sheathing.
- Similar cost as standard truss roof.
- Truss roof typically has lower material and installation costs than other roof framing options.
- Truss roofs can use smaller dimension lumber than other roof types.

 ***Disadvantages:***

- Flat soffit option requires additional framing.

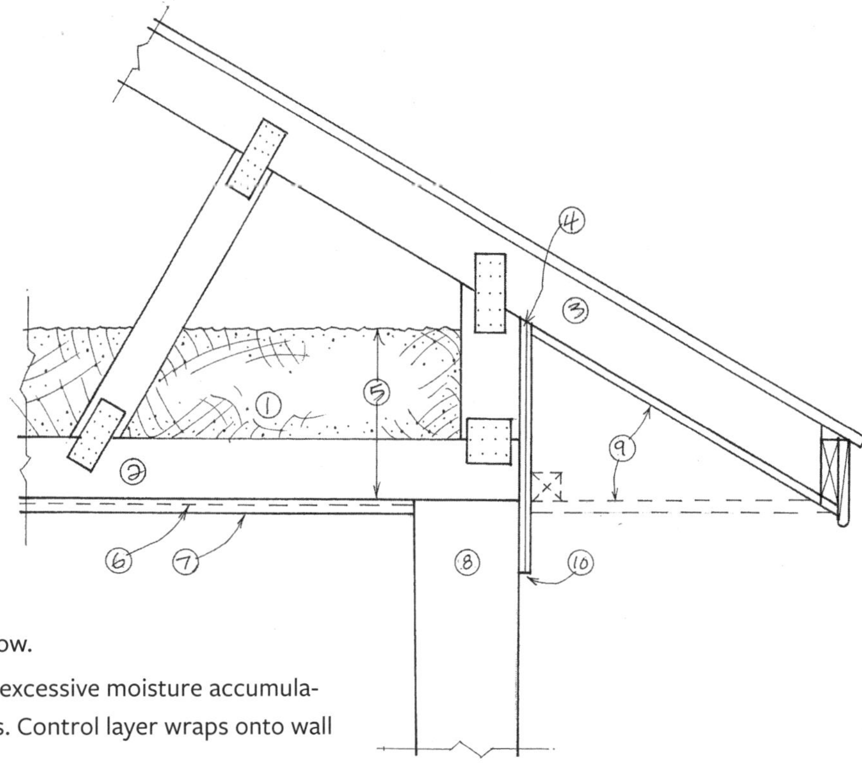

1. Hempcrete insulation @ 200 kg/m³
2. Truss bottom chord, designed to accommodate dead load of hempcrete
3. Truss top chord, designed for appropriate loads. Bottom chord ends at outside edge of wall.
4. Insulated sheathing acts as insulation baffle, may be continuous with wall sheathing. Provides full depth of truss top chord for ventilation.
5. Full depth of insulation must be reached directly over the wall to meet thermal performance requirements. Roof pitch can be shallow.
6. Vapor control layer below insulation to prevent excessive moisture accumulation in attic area, as per local code requirements. Control layer wraps onto wall to provide air sealing at junction.
7. Finished ceiling, able to handle dead load of hempcrete. Insulated sheathing could be applied prior to the ceiling.
8. Hempcrete wall of appropriate design and thickness
9. Ventilated soffit, as per local code requirements. Soffit can be fastened to underside of truss top chord, or can be framed to a horizontal position.
10. Insulated sheathing prevents thermal bridging of truss heel, and can be continuous with wall sheathing to provide air sealing at truss/wall junction

Truss or I-Beam Vaulted Roof

A vaulted roof space can provide additional habitable space within the same building footprint. Some form of manufactured roof framing needs to be used to provide adequate insulation depth, as solid framing lumber cannot be purchased in a suitable depth for most climates.

 ***Advantages:***

- Provides vaulted roof space inside building.
- Single framing member provides adequate insulation depth.
- Less material than truss roof.

 ***Disadvantages:***

- Deep framing members required for adequate thermal performance.
- Thermal bridging from roof framing can be difficult to minimize.
- Bearing point must be properly designed.
- Insulation baffles must be inserted between each framing member.
- Soffit design and detailing must be properly designed and executed.

1. Hempcrete insulation @ 200 kg/m^3
2. Truss or I-beam bottom chord, designed to accommodate dead load of hempcrete
3. Truss or I-beam top chord, designed for appropriate loads.
4. Insulation baffle inserted between roof framing members with appropriate ventilation space as per local codes
5. Full depth of insulation must be reached directly over the wall to meet thermal performance requirements.
6. Vapor control layer below insulation to prevent excessive moisture accumulation in attic area, as per local code requirements. Control layer wraps onto wall to provide air sealing at junction.
7. Finished ceiling, able to handle dead load of hempcrete. Insulated sheathing could be applied prior to the ceiling.
8. Hempcrete wall of appropriate design and thickness
9. Ventilated soffit, as per local code requirements. Soffit is typically framed to a horizontal position

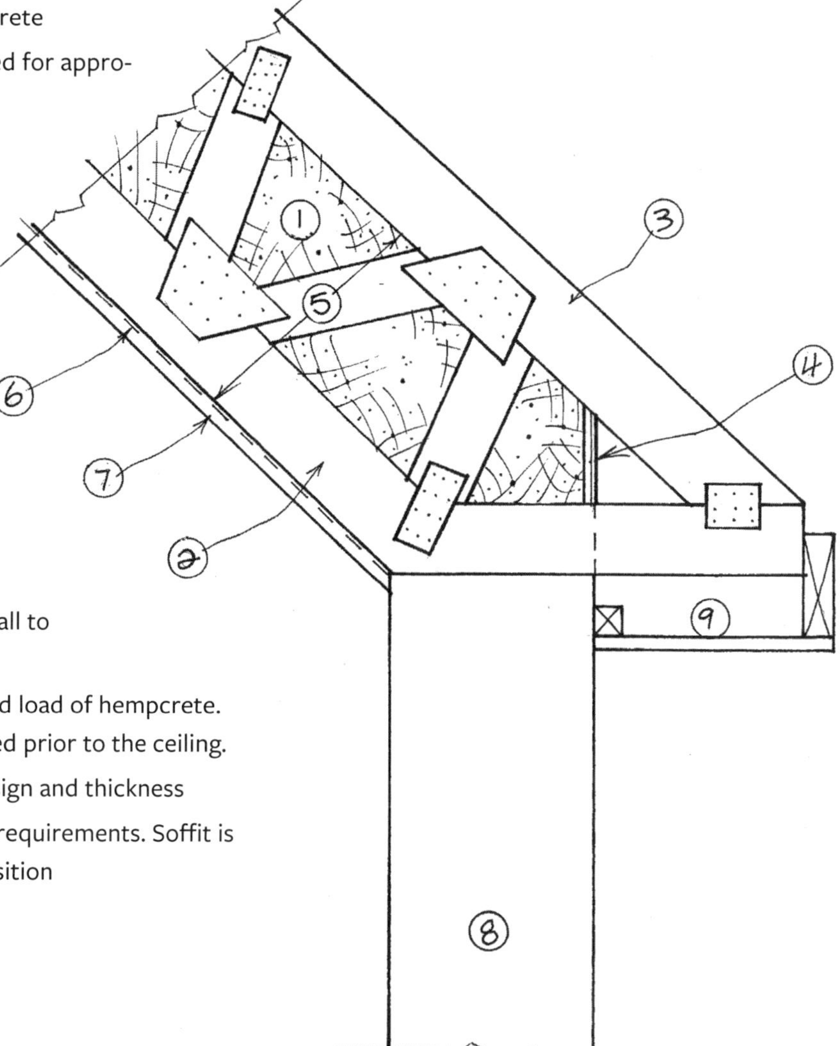

Window/Door Insulation

The use of hempcrete as an insulation material around windows is not supported by any testing documents, but has been used in northern climates with monitored success. The material has many properties that make it suitable for this purpose:

- It does not shrink when curing, keeping tight contact with framing and window units.
- It does not swell or shrink appreciably with temperature fluctuations.
- It maintains some degree of flexibility.
- It is able to fill irregular voids.
- It can be cut and removed easily when windows need replacement.

The mix ratios for window insulation are unique from other forms of hempcrete (see Mixes chapter), with the inclusion of hemp fiber and more binder and water in the mix.

Hempcrete is packed into the shim space of a window unit, then hand formed to create an exterior (and/or interior) trim profile that seals the window and insulates the framing. The final trim can be plastered and/or painted.

Hempcrete window insulation should not be relied upon to provide a completely airtight seal. To ensure a high degree of air tightness there are two tactics that can work:

- Fill the gap between the window unit and framing with hempcrete and use an appropriate air sealing tape adhered to the window unit and the framing on at least one side of the window, or preferably both sides.
- Continue the hempcrete window insulation to one or preferably both sides of the window so that it wraps onto the face of the window frame and is built into a trim.

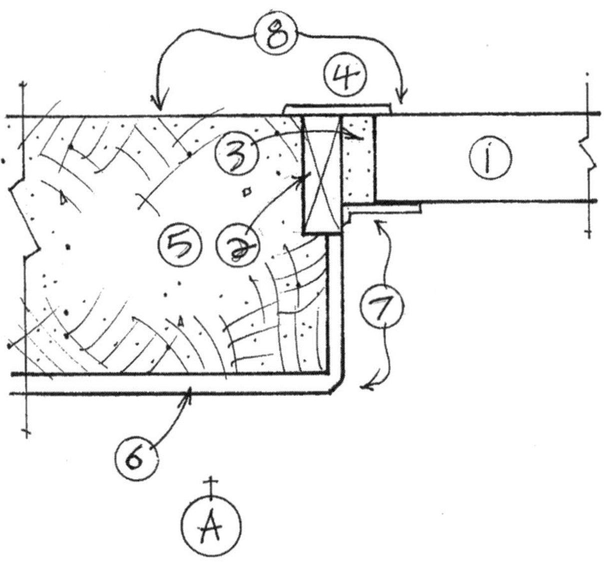

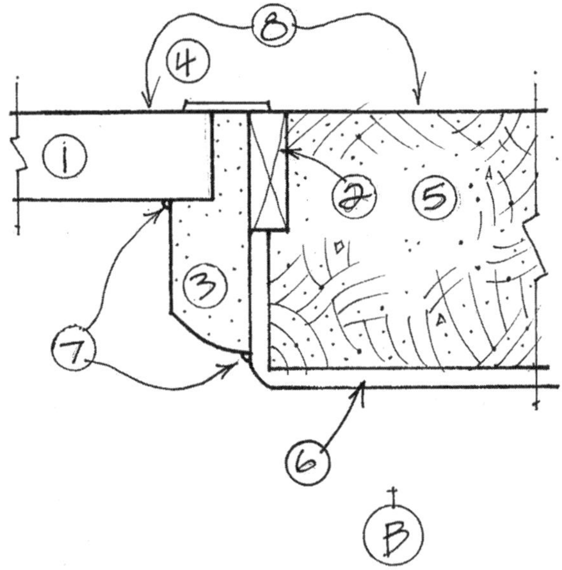

A) ***Tape and Conventional Trim***

Hempcrete window insulation is used only between the window unit and the framing, and an air seal is created with an appropriate sealing tape on both sides.

1. Window unit. Details will vary with window, including depth of unit, jamb extensions, flanges and sills. Consult window details prior to specifying a particular approach for hempcrete insulation.
2. Framing for window. Details will vary with placement within the wall thickness, size of framing and intended finishes.
3. Hempcrete window insulation, lightly packed to ensure entire void is filled
4. Tape for window sealing. Ensure that tape is designed to affix to window frame material and framing.
5. Hempcrete wall insulation
6. Interior finish (plaster, wallboard or other)
7. Interior trim detail to suit window type, framing, wall depth and aesthetic needs

B) ***Tape on Exterior, Hempcrete Trim Interior***

Hempcrete window insulation is used to insulate shim space and to form a trim profile on the interior face. An appropriate tape is used to create an exterior seal.

1. Window unit. Details will vary with window, including depth of unit, jamb extensions, flanges and sills. Consult window details prior to specifying a particular approach for hempcrete insulation.
2. Framing for window. Details will vary with placement within the wall thickness, size of framing and intended finishes.
3. Hempcrete window insulation, lightly packed to ensure entire void is filled. Interior trim profile can be formed with temporary formwork or shaped by hand.
4. Tape for window sealing. Ensure that tape is designed to affix to window frame material and framing.
5. Hempcrete wall insulation
6. Interior finish (plaster, wallboard or other)
7. Caulking may be applied at edges of dried hempcrete trim to ensure air tightness
8. Exterior finish and trim as per design requirements

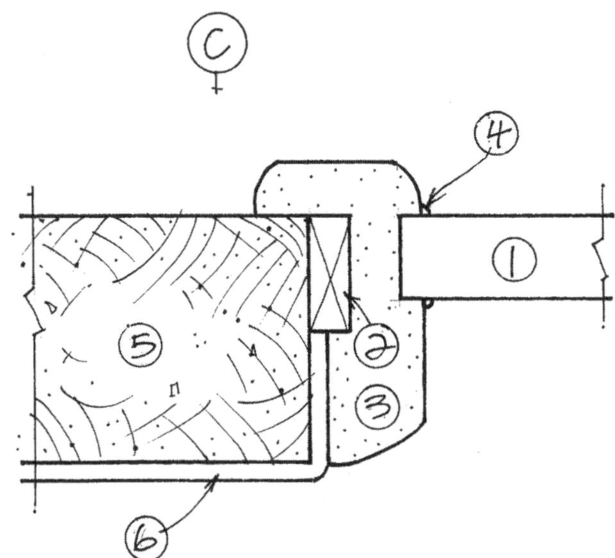

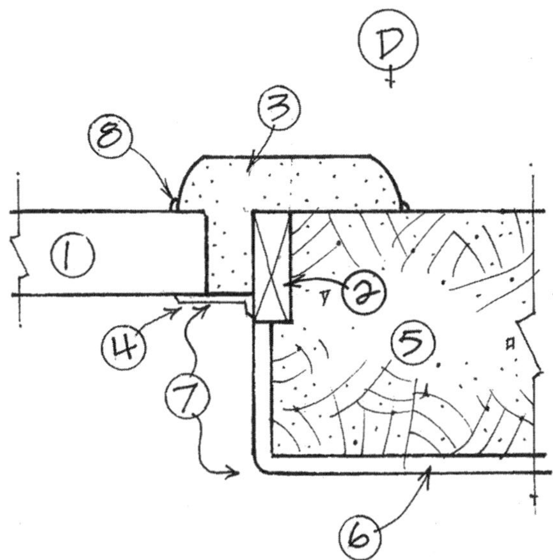

C) ***Hempcrete Trim Interior and Exterior***
Hempcrete window insulation is used to insulate shim space and to form a trim profile on the interior and exterior face. Insulation surrounding framing can lower thermal bridging of window framing.

1. Window unit. Details will vary with window, including depth of unit, jamb extensions, flanges and sills. Consult window details prior to specifying a particular approach for hempcrete insulation.
2. Framing for window. Details will vary with placement within the wall thickness, size of framing and intended finishes.
3. Hempcrete window insulation, lightly packed to ensure entire void is filled. Interior and exterior trim profile can be formed with temporary formwork or shaped by hand.
4. Caulking for air and water sealing
5. Hempcrete wall insulation
6. Interior finish (plaster, wallboard or other)

D) ***Hempcrete Trim Exterior, Tape on Interior***
Hempcrete window insulation is used to insulate shim space and to form a trim profile on the exterior face. Insulation surrounding framing can lower thermal bridging of window framing.

1. Window unit. Details will vary with window, including depth of unit, jamb extensions, flanges and sills. Consult window details prior to specifying a particular approach for hempcrete insulation.
2. Framing for window. Details will vary with placement within the wall thickness, size of framing and intended finishes.
3. Hempcrete window insulation, lightly packed to ensure entire void is filled. Exterior trim profile can be formed with temporary formwork or shaped by hand.
4. Tape for window sealing. Ensure that tape is designed to affix to window frame material and framing.
5. Hempcrete wall insulation
6. Interior finish (plaster, wallboard or other)
7. Interior trim detail to suit window type, framing, wall depth and aesthetic needs
8. Caulking for air and water sealing

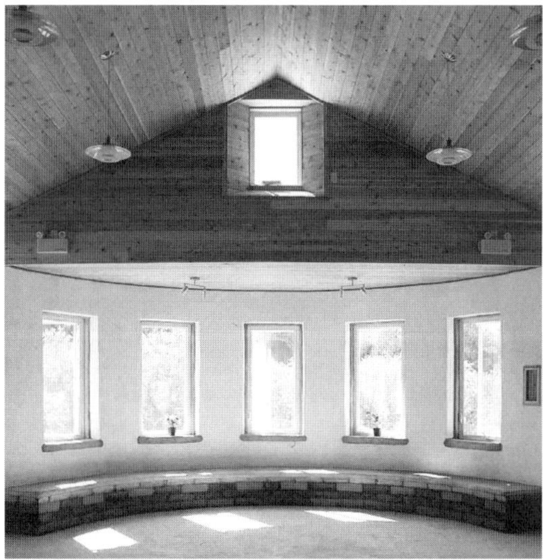

Hempcrete trim can easily be formed to any shape, and makes attractive exterior trim. On the interior, hempcrete can be trim and windowsill material (and bench top as well).

Chapter 7

Design Notes for Hempcrete

A designer of buildings using hempcrete insulation will need to choose from the wide range of options presented in the previous chapter in order to create a building that functions as an integrated system.

Specification of all connections, intersections and layers requires understanding of more than just the hempcrete insulation. A designer should consider all the points given on the following two pages.

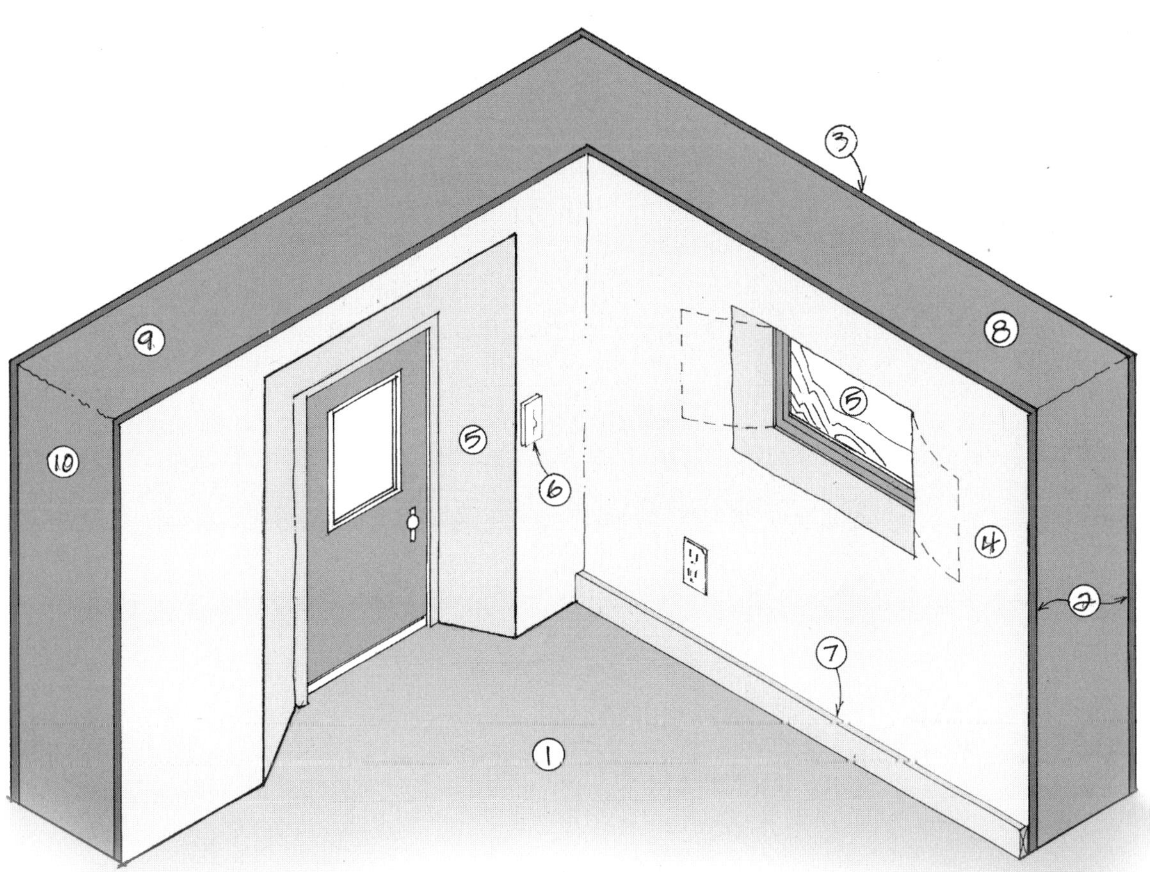

1) *Foundation*

a) Wood frame walls for hempcrete are adaptable to any building foundation system. Connection types will be specific to foundation materials and design and should follow local code requirements.

b) Provide the walls a reasonable elevation above grade to prevent precipitation overloading at the base of the wall and to keep the wall above expected snow depths.

c) Provide adequate foundation floor drainage to ensure minor flooding does not overwhelm walls.

d) Overall wall width may require blocking in wood frame foundations under the inner face of the wall.

e) It may be possible to cantilever the wall system beyond the edge of the foundation to reduce the impact of wider walls on the foundation size.

f) Provide details for effective flashing/sealing detail at wall/foundation intersection.

g) Specify a suitable moisture break layer between foundation and wall.

h) Specify a suitable air barrier to prevent leakage between wall and foundation. Caulking, adhesive, expanding tape and/or surface tape should be chosen to suit site and code conditions.

2) *Overall wall width*

a) Wall thickness should be designed so hempcrete insulation meets thermal performance targets.

b) Overall wall thickness must account for hempcrete insulation depth plus all finishing layers, including plaster and/or sheathing and any furring associated with siding/sheathing systems.

c) Thick layers of insulated sheathing included to increase thermal performance must be attached in an appropriate manner.

3) *Exterior finishes and wall protection*

a) Specify roof overhang and gutters appropriate for climate to protect walls from excessive precipitation.

b) Specify finishes appropriate to climate and exterior conditions, including a minimum permeability for exterior finishes.

c) Where codes and/or performance targets require the use of a sheet-style barrier on the exterior of the building, specify the highest perm rating for such barriers and be sure that the barrier can be adequately fastened to the wall.

d) Rain screen siding should be applied over a furring space of at least ½ inch (12.5 mm), and air must be able to circulate freely in vertical channels protected from insect/pest intrusion.

4) *Interior finishes*

a) Proper air sealing details must be included for all wall seams, intersections and penetrations.

b) Plaster finishes should be minimum two-coat systems, and mix ratios and application thickness should be specified. Where plaster covers hempcrete and wood, wood surfaces must be meshed or treated to prevent cracking.

c) Wallboard finishes must be installed tightly against hempcrete insulation, and fastening points should meet manufacturer specifications.

d) The use of Class I or II vapor retarders should be avoided in the wall system.

e) Perm ratings of interior finishes must be considered in all layers, including paints/surface treatments.

5) *Window and door details*

a) Placement of windows/doors in the depth of the wall will determine sill requirements and return/trim details.

b) Provide positive drainage and proper drip edge on exterior sill.
c) Provide adequate exterior flashing at top edge of windows/doors.
d) Rounded, flared or angled window openings will require appropriate framing details.
e) Hempcrete window insulation and/or trim should be tested for air tightness.

6) ***Electrical details***

a) Any electrical wiring encased within hempcrete insulation should be shielded from contact with the hempcrete in a manner that meets local electrical codes for wiring in contact with concrete.
b) All electrical boxes must be detailed to prevent air leakage into the wall assembly.
c) Electrical boxes should be connected to framing members and not anchored in hempcrete.
d) Electrical box connection details should take into account the full depth of all intended finishes to be applied to hempcrete insulation.

7) ***Attachment points***

a) Framing should be provided for the attachment of all trim and other permanent installations. Do not anchor directly to hempcrete insulation.

8) ***Top of wall***

a) Provide details for appropriate fastening of roof and floor framing to structural frame.
b) Specify treatment of insulation at top of wall. If sealing is required, lime plaster should be specified.
c) Where hempcrete roof insulation is also being used, provide key at top of wall to ensure adequate junction between wall and roof insulation.

9) ***Hempcrete specifications***

a) All relevant details for hempcrete insulation should be included in plans, including specifications for hemp hurd and lime binder, mix ratios, and intended density.
b) Specify the formation of test cylinders of hempcrete mix to ensure density and other properties are met.
c) Include drawing sheet with design and specification of intended formwork/shuttering and attachment to frame.
d) Curing and drying time for hempcrete wall insulation must be specified. A meter that reads wood moisture equivalent should indicate moisture content of no more than 25% at the center of the wall before the hempcrete is enclosed on both sides. Depending upon weather conditions this can take from 2 to 6 weeks. In rare cases (cold damp weather, a mix ratio that was too wet) drying can take up to 8 weeks.

   The use of an industrial dehumidifier and fans can significantly speed up the drying process, and can be included in the construction notes.

Designed by Chris and Will Dancey. Built in 2005, Aylmer, Ontario, Canada. Hempcrete installation by ArtCan/ Gabriel Gauthier

Hempcrete walls are combined with a timber frame for this house that also combines round and straight wall sections. It's the hub of activity at Dancey Family Farm. Credit: ArtCan/ Gabriel Gauthier

Designed by Alex and Marylene. Built in 2011, in the Eastern Townships, Quebec, Canada, with support from Jean-Louis Valentin and ArtCan/Gabriel Gauthier.

The owners of this home in Quebec modeled the timber frame on a traditional design from the Champagne region in France, where modern hempcrete was first used. The hempcrete insulates well against the cold northern climate. CREDIT: ARTCAN/GABRIEL GAUTHIER

Designed by Alex and Marylene. Built in 2011, in the Eastern Townships, Quebec, Canada, with support from Jean-Louis Valentin and ArtCan/Gabriel Gauthier.

Bright white lime plaster on hempcrete is a perfect contrast for wood surfaces, and helps to make a large space bright and welcoming CREDIT: ARTCAN/GABRIEL GAUTHIER

Designed by Alex and Marylene. Built in 2011, in the Eastern Townships, Quebec, Canada, with support from Jean-Louis Valentin and ArtCan/Gabriel Gauthier.

The combination of large timbers and hempcrete creates an old European aesthetic. Hempcrete was "invented" as a means of replacing old mud/straw insulation in historic timber frame buildings. CREDIT: ARTCAN/GABRIEL GAUTHIER

Right: Designed and built by the owner Patrick Connor with support from ArtCan/Gabriel Gauthier. Built in 2009, near Chelsea, Quebec.

This owner-designed and -built home uses design cues from traditional Quebec homes and incorporates hempcrete wall insulation and lime plaster into this aesthetic. CREDIT: ARTCAN/GABRIEL GAUTHIER

The Nauhaus Prototype, designed by Alembic Studio. Built in Asheville, North Carolina

Above: *Given the temperate climate of the region, every room in this home, with the exception of one lower level bedroom, is designed with access to exterior living areas. In this view, the porches in the front of this home provide expanded areas for lounging and dining.* Credit: Anthony Abraira

Center left: *Southwest corner of Nauhaus Prototype; the longer axis of the South elevation exhibits window openings calculated for varying degrees of solar heat gain, based on the Passive House Planning Package — only one of the components for achieving exemplary energy efficiency. The porches on the West provide outdoor living connections to the adjacent bedrooms.* Credit: Anthony Abraira

Below left: *This space is well-lit and expansive owing to the abundance of light transmitted through the windows and wall openings.* Credit: Anthony Abraira

Above: Designed by Chris Magwood. Built in 2004 near Madoc, Ontario, Canada.

This house was built with all-local materials, including hemp grown just down the road. Hempcrete window insulation and window sills are combined with hemp straw bales and clay plaster with hemp hurd and fibers, and was the author's first use of hempcrete. Credit: Ben Bowman

Center right: Designed by Chris Magwood and Soma Earth Architect. Built in 2008, Madoc, Ontario Canada.

Arts Centre Hastings uses locally grown hemp in almost every way possible, including hempcrete wall, floor, foundation and window insulation, as well as window trim. Bales of hemp make up some of the walls, and the clay plaster used inside and out also uses hemp hurd and fiber. Credit: Ben Bowman

Bottom right: Designed by Jen Feigin and Chris Magwood. Built in 2011, Jacob Island, Ennismore, Ontario, Canada.

The Camp Maple Leaf family cabin features a range of natural building materials, including hempcrete wall insulation and natural clay and lime plasters using hemp fiber and hurd. Credit: Jen Feigin

Designed by Jen Feigin, Chris Magwood and Our Cool Blue Architects. Built in 2009, Peterborough, Ontario, Canada

The Camp Kawartha Environment Centre is a leading teacher of sustainability, offering programs for children and adults. The building features hempcrete wall insulation on the southern exposure of its passive solar design, and hempcrete window trim throughout. Credit: Daniel Earle

The cozy southern section of the Camp Kawartha Environment Centre features hempcrete wall insulation, hempcrete window trim and sills, a hempcrete bench top and hemp in the lime plaster. Credit: Chris Magwood

Hempcrete can be used to insulate around windows, and formed to create raised trim on the exterior and interior of a building. Painted with natural casein paint, the hempcrete can achieve straight or rounded shapes easily, offering a wide range of aesthetic options. Credit: Chris Magwood

Mandala Home built with hempcrete walls in 2012, Nelson BC. Credit: Jayeson Henderson/ Hempcrete Natural Building

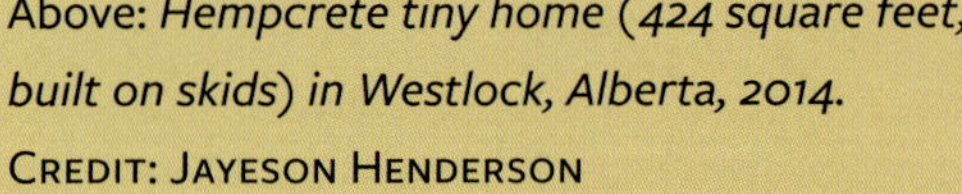

Above: *Hempcrete tiny home (424 square feet, built on skids) in Westlock, Alberta, 2014.*
Credit: Jayeson Henderson

Center right: *Exterior of Hempcrete Cottage, 2008, with hempcrete addition and hempcrete garden wall to the left.*
Credit: Jayeson Henderson/Hempcrete Natural Building

North America's first hempcrete kitchen, built in 2003 on Bowen Island, B.C.
Credit: Jayeson Henderson

Above: Designed by Aukett Swanke. Built near Liverpool, England in 2012.

The Marks & Spencer Cheshire Oaks store is a great example of a more sustainable approach to large buildings, and includes hempcrete wall insulation among a host of other sustainable features built into its 210,000 square foot design. It's an aspirational example for North American builders. Credit: Paul White

Center left: Built in 2012, Tarpon Springs, Florida.

Owner-builder Bob Clayton used hempcrete wall and attic insulation to create the Tarpon Springs Hempstead, putting the material to use in a hot, humid climate. Credit: Bob Clayton

Bottom left: *The first approved use of hempcrete in California was for an addition to a ranch-style house in Rolling Hills. The Galvin/Kirmse building garnered a lot of attention during the 2014 build, and has opened the way for many projects now in the planning stages in the state.* Credit: Beate Kirmse

Designed by Alembic Studio. Built in Virginia Beach, Virginia

Mare Cannabum, constructed on the mid-Atlantic coast. This hempcrete home was engineered to withstand hurricane force winds. The modest second floor and gable cantilevers, reminiscent of traditional European designs, provide protection for the walls below from typical rain events. CREDIT: JENNIFER BENNETT

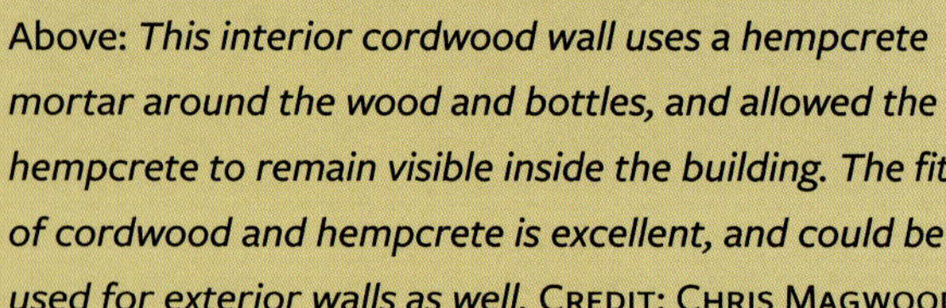

Above: *This interior cordwood wall uses a hempcrete mortar around the wood and bottles, and allowed the hempcrete to remain visible inside the building. The fit of cordwood and hempcrete is excellent, and could be used for exterior walls as well.* CREDIT: CHRIS MAGWOOD

Center right: *Designed by Alembic Studio. Built in Asheville, North Carolina.*

Tim Callahan inspects the installation in this newly constructed hempcrete wall. The door on the left has been formed up with chamfered jambs. Electrical boxes are cast in place and connected by conduit. CREDIT: JENNIFER BENNETT

Hempcrete is used to create bottle transoms over the doorways in an office building. This is a great way to leave some hempcrete visible, and to make use of the sculptable nature of hempcrete. CREDIT: CHRIS MAGWOOD

Chapter 8

Hempcrete Mixes

Hempcrete has a number of variables that can make it difficult to predict exact results, even if the same proportions of hemp hurd, lime binder and water are used. Variables include:

- **Particle size distribution of hemp hurd.** A higher percentage of small particles will have more surface area, requiring the binder to cover more area and resulting in less binder coating the hemp. More small particles will also mean less air space between particles, as they will fit together more tightly. The opposite will happen if there is a higher percentage of large hemp particles.
- **Type of lime binder.** As will be shown in the mix recipes that follow, there are several types of lime binder that are viable for hempcrete mixes. They will all have different properties, including the rate at which the binder ingredients mix with water and distribute over the hemp particles, the speed of the hydraulic set, and the actual density of the binder portion.
- **Amount of water in the mix.** Small variations in water quantity can have a large impact on final density of the mix, as well as the quality of the hydraulic set. Even if water quantities are measured accurately, variations in the humidity of the hemp hurd can have a noticeable impact, as can slight variations in volume of binder and hemp with the same amount of water.
- **Mixer type and amount of mixing.** The various types of mixing machinery can generate hempcrete with different properties, even if all the ingredient ratios are the same. And

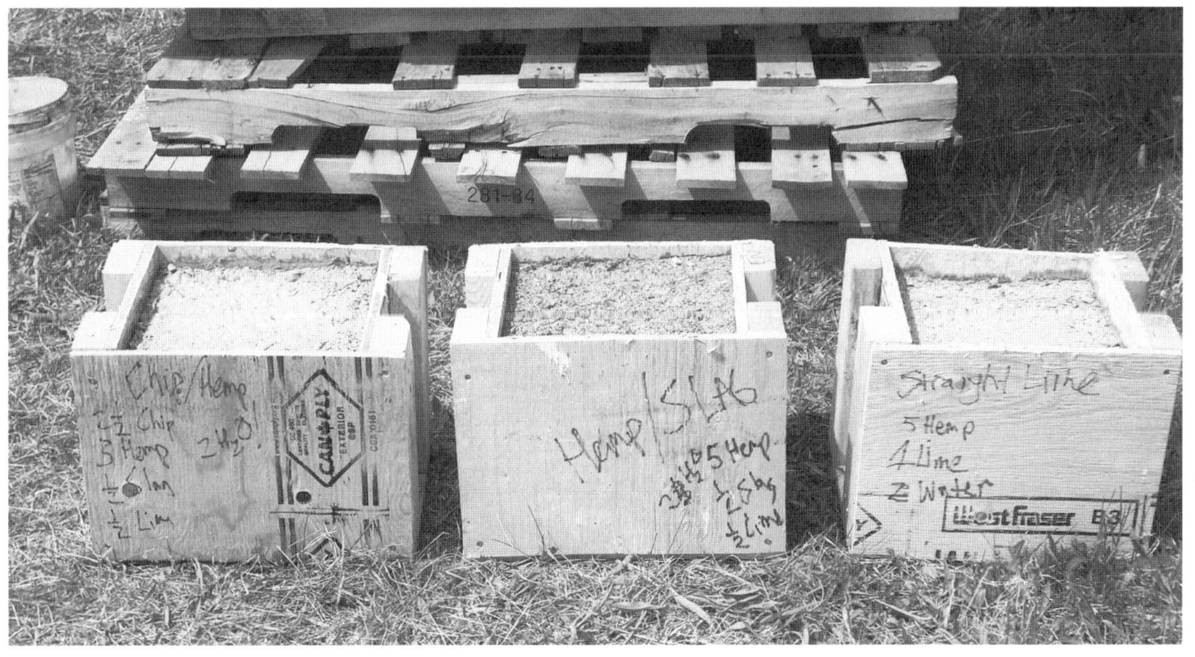

Form boxes with a volume of one cubic foot can help to ensure proper mix ratios — and that different workers are using consistent tamping procedures.

within the same mixing machine, the length of time the ingredients spend in the mixer can affect the resulting hempcrete.

- **Amount of tamping.** This is likely to be the largest variable, and the most difficult to control. With the exception of loose fill applications in attics, some degree of tamping is applied to hempcrete to ensure that it completely fills cavities and achieves a suitable amount of contact between particles to give a desired amount of cohesion. This light compaction is done by hand (see Construction Procedure chapter), or using a simple wooden tamper. The exact same mix can be tamped such that the density can double, and it can be very difficult to provide training and quality control to ensure a consistent compaction.

With all these variables in mind, it is important that the mix ratios described here be considered as starting points. Samples should be made and tested. The samples should use the same ingredient sources and be produced in the same mixing machinery that will be used for the building. If workers are new to installing hempcrete, they should create a sample block and be shown how to do the correct amount of tamping. Samples from each worker can be dried and weighed, and the results can help to inform the process.

Despite all the variables, it's not that difficult to achieve a viable consistency in the hempcrete insulation. The mix recipes that follow will put you in the right ballpark, and conscientious mixing and installation will ensure the results meet expectations.

Hemp-to-binder Ratios

Mix ratios for hempcrete wall insulation are a balance between density that is low enough to provide the best thermal performance with density that is high enough to provide adequate structural characteristics. Different approaches to wall design will affect the type of mixture used.

Table 8.1: Mix ratios for hempcrete insulation

Mix Type	Type of Insulation	Tamping Required	Hemp Hurd (by weight)	Lime Binder* (by weight)	Water** (by weight)
Very Light Density 9.4 to 12.5 lb/ft^3 (150 to 200 kg/m^3)	Loose fill, flat ceilings and floor joist cavities	None	1	1	1.5
Light Density 12.5 to 15.6 lb/ft^3 (200 to 250 kg/m^3)	Sloped attic spaces, walls with structural sheathing on both sides	Gentle, minimal compression	1	1 to 1.5	1.5 to 2
Medium Density 15.6 to 21.8 lb/ft^3 (250 to 350 kg/m^3)	Walls, with plaster finish or nonstructural sheathing	Light compression	1	1.5 to 2	1.5 to 2
High Density 21.8 to 31.2 lb/ft^3 (350 to 500 kg/m^3)	Sub-slab insulation	Firm compression	1	2 to 3	2 to 2.5

* *Weight is total of all binders for multi-component binder recipes. See* Binder Ingredients and Formulas. *For proprietary binders, follow the manufacturer's instructions.*
** *Water volume must be tested and adjusted. Final volume may not fall within this range.*

Very Light Density Mixes — in the range of 9.4 to 12.5 lb/ft^3 (150 to 200 kg/m^3) — are suitable for loose fill applications where the hempcrete is not required to have any cohesion or structural properties. Flat attic spaces, floor joist systems and fully sheathed walls are examples of places where very light mixes would be applicable. These mixes are not tamped at all during installation.

1. Very light density hempcrete, un-tamped. Roof framing and ceiling capable of handling dead load of 9.4 to 12.5 lb/ft^3 (150 to 200 kg/m^3).
2. Very light density hempcrete, un-tamped. Floor framing and underside sheathing capable of handling dead load of 9.4 to 12.5 lb/ft^3 (150 to 200 kg/m^3).

Light Density Mixes — in the range of 12.5 to 15.6 lb/ft^3 (200 to 250 kg/m^3) — are suitable for wall systems that incorporate a sheathing material on both sides of the wall. In such a wall system, the hempcrete does not have to provide much meaningful cohesion since the structural sheathing is providing a solid wall surface to attach finishes and providing the wall framing with adequate shear strength and resistance to bending of framing members. In such wall systems, one side of the wall is typically sheathed prior to hempcrete installation, providing a permanent formwork on one side of the wall. The hempcrete only needs enough cohesion to stay in place when the form on the other side of the wall is removed to allow the hempcrete to dry. Once the hempcrete is enclosed within the sheathing, it does not need to be cohesive.

This type of mix is also suitable for roof insulation in sloped roof cavities, where a small amount of cohesion keeps the hempcrete insulation from slumping or sliding down the slope due to gravity.

This type of mix is tamped into place; just enough force is used to ensure that hurd particles are making contact with one another.

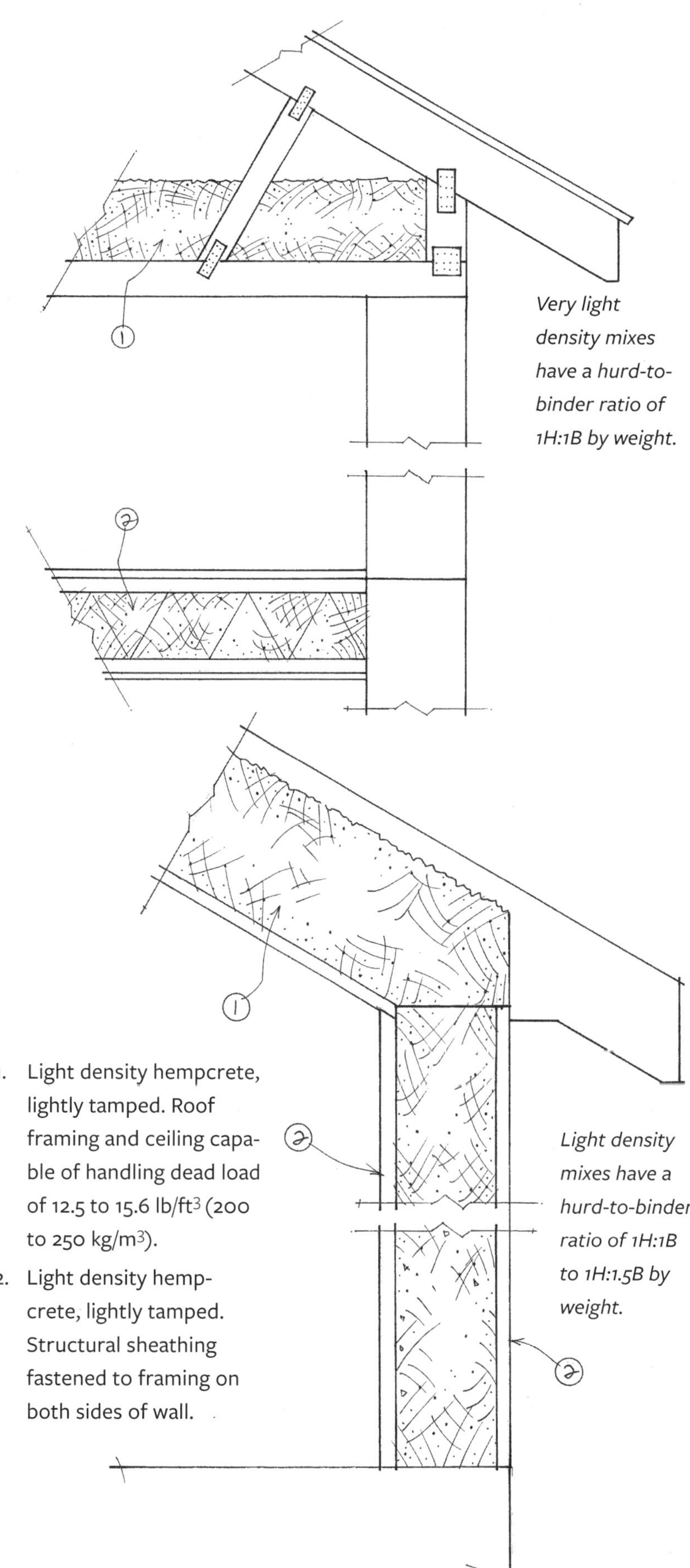

Very light density mixes have a hurd-to-binder ratio of 1H:1B by weight.

1. Light density hempcrete, lightly tamped. Roof framing and ceiling capable of handling dead load of 12.5 to 15.6 lb/ft^3 (200 to 250 kg/m^3).
2. Light density hempcrete, lightly tamped. Structural sheathing fastened to framing on both sides of wall.

Light density mixes have a hurd-to-binder ratio of 1H:1B to 1H:1.5B by weight.

Medium Density Mixes — in the range of 15.6 to 21.8 lb/ft^3 (250 to 350 kg/m^3) — are suitable for wall systems in which one or both sides of the wall will be finished with a direct-applied plaster. The plaster finish requires that the hempcrete have enough cohesion that the plaster will grip the surface and that the hempcrete will not crumble or dislodge and allow the plaster to delaminate from the surface. Experience with medium density wall mixes has shown them to be a very reliable substrate for plaster. Medium wall mixes can also be used when one side of the wall has structural sheathing and the other side has a plaster finish.

These mixes are tamped enough to provide a small degree of compaction, forcing the hemp particles to settle into one another and provide plenty of surface contact without eliminating open pore spaces between particles.

High Density Mixes — in the range of 350 to 500 kg/m^3 (21.8 to 31.2 lb/ft^3) — are used as sub-slab insulation where the hempcrete needs to be able to bear the compressive load of the slab floor. A structural engineer should ensure that the bearing capacity of the hempcrete insulation is suitable. These mixes are tamped firmly, forcing the hurd particles into close contact with one another.

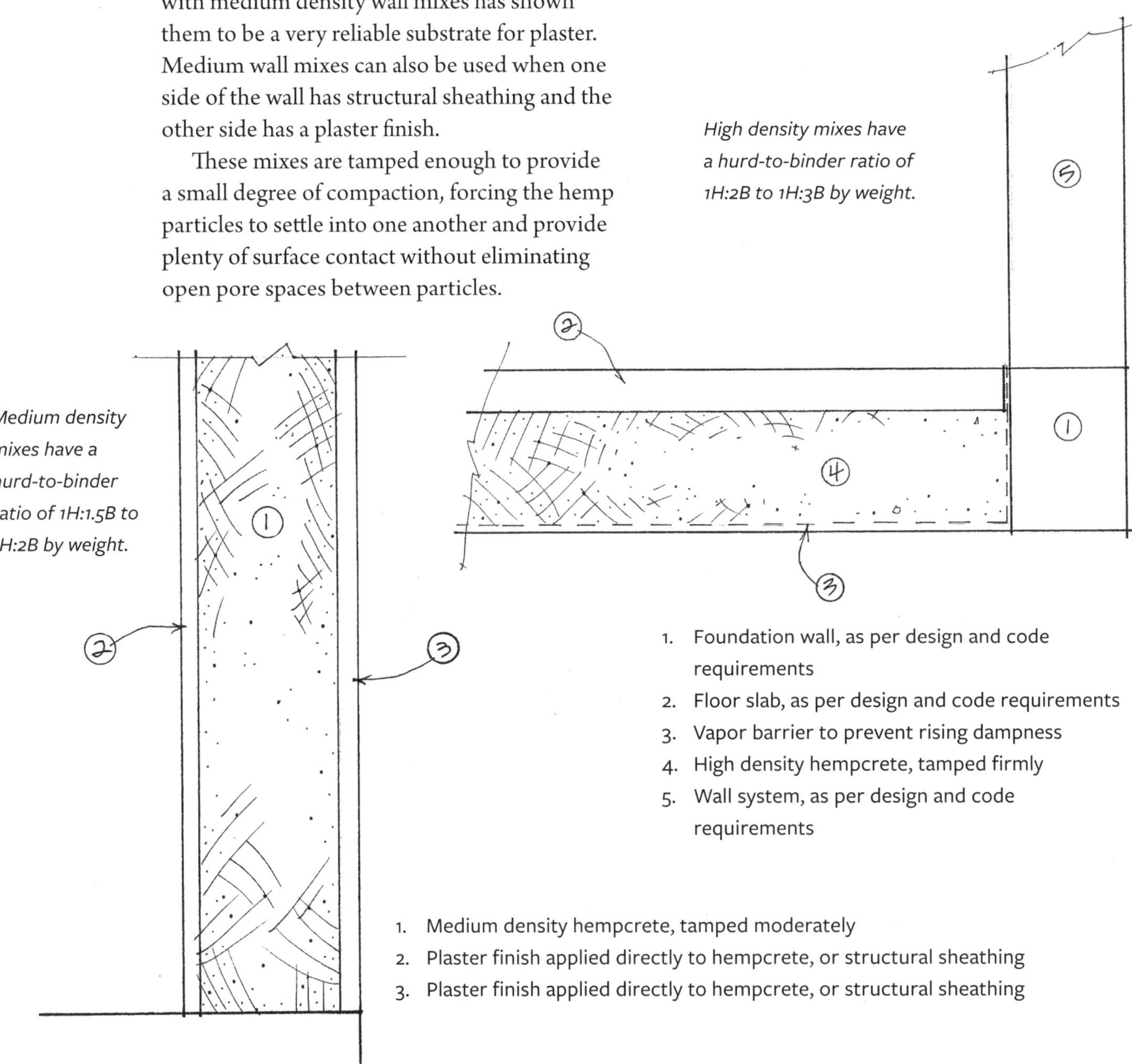

High density mixes have a hurd-to-binder ratio of 1H:2B to 1H:3B by weight.

1. Foundation wall, as per design and code requirements
2. Floor slab, as per design and code requirements
3. Vapor barrier to prevent rising dampness
4. High density hempcrete, tamped firmly
5. Wall system, as per design and code requirements

Medium density mixes have a hurd-to-binder ratio of 1H:1.5B to 1H:2B by weight.

1. Medium density hempcrete, tamped moderately
2. Plaster finish applied directly to hempcrete, or structural sheathing
3. Plaster finish applied directly to hempcrete, or structural sheathing

Binder Ingredients and Formulas

The binders used for hempcrete can have many formulations. They all contain hydrated lime, but there are a variety of admixtures used to create a hydraulic set. A binder recipe of 100% hydrated lime should not be used because the curing of hydrated lime is by carbonization; this is a slow process that will result in very long drying/setting times and likely to an incomplete setting in the middle of the volume where the lime does not have much access to carbon dioxide with which to react.

Hydraulic setting agents for hempcrete binders come in two varieties: *hydraulic* and *pozzolanic.*

Hydraulic binders include natural hydraulic lime (NHL) and natural cement, Portland cement, and magnesium cement. These ingredients have an independent hydraulic action when exposed to water and do not require the presence of hydrated lime. Hydraulic binders are typically mixed with hydrated lime at ratios of 20–50% by volume. While it is possible to bind hempcrete only with a hydraulic binder and no hydrated lime, this is not typically done because hydrated lime is less expensive, has a lower density and contributes excellent moisture handling abilities and antifungal and antimicrobial qualities to the mix. Note that mixes with more than 10% Portland cement are considerably denser and have much lower moisture storage capabilities.

Pozzolanic binders include metakaolin (calcined kaolin), fly ash, blast furnace slag and zeolite. These ingredients only have a hydraulic action when combined with hydrated lime and water. Pozzolanic binders are typically mixed with hydrated lime at ratios of 40–50% by volume. These binders are inert without the presence of hydrated lime and will have no binding effect if used without hydrated lime. Metakaolin is the only pozzolanic binder that has been tested for hempcrete; the remaining pozzolans are likely viable but would require experimentation and testing to confirm ratios and properties.

Table 8.2: Binder mix ratios

Binder Ingredients	Hydrated Lime Ratio (by volume)	Additive Ratio (by volume)	Properties
Hydrated lime & NHL 5 (natural hydraulic lime) or natural cement	50-80%	20-50%	Well-tested; excellent moisture handling; expensive; more difficult to source
Hydrated lime & metakaolin	50-60%	40-50%	Fewer tests; excellent moisture handling; least expensive; more difficult to source
Hydrated lime & magnesium cement	25-75%	25-75%	Fewer tests; moisture handling not quantified; more expensive; more difficult to source
Hydrated lime & Portland cement	75-90%	10-25%	Less desirable moisture handling properties; creates denser mix; less expensive; easy to source

Water Ratio

The ratio of water used in a hempcrete mixture is critical. The right ratio of hemp hurd to binder can be viable or be ruined by the wrong amount of water.

Too much water will saturate the hemp hurds, which are extremely porous and capable of absorbing a high volume of water. This can affect the dispersion of the binder over the hemp hurd and the curing of the binder, resulting in a weaker mix. Excess water can also greatly prolong the drying process for the wall, which can be highly problematic. Extended drying times not only affect the construction schedule of the project (walls cannot be enclosed/plastered until they are dry) but walls that remain too wet for too long can develop mold, despite the presence of the lime. Framing and other building components exposed to very wet hempcrete for extended periods of time can also be negatively affected.

Too little water will result in a mixture that does not have adequate distribution of moist binder over the surface of the hemp and not enough water to fully achieve the hydraulic chemical setting of the binder. This will result in a mixture that will not tamp to become cohesive, creating a weak and crumbly mixture.

The right amount of water will require some experimentation to get right. Adding *1.5 to 2 parts of water (by weight) for every 1 part of hemp hurd* is a good starting point for experimentation. We have had viable mixes that use both less and more water than this, but it will be easy to adjust from this point.

When experimenting with water ratios, a simple test is to use gloved hands to form some of the mix into a ball. If the ball can be squeezed by hand into a coherent shape that maintains its integrity, the water ratio is in the right range.

For very light or light density mixes, the squeezed ball should be quite fragile; it will keep

A few chances to handle different mixes will quickly give installers a good sense of the proper consistency of hempcrete mixes.

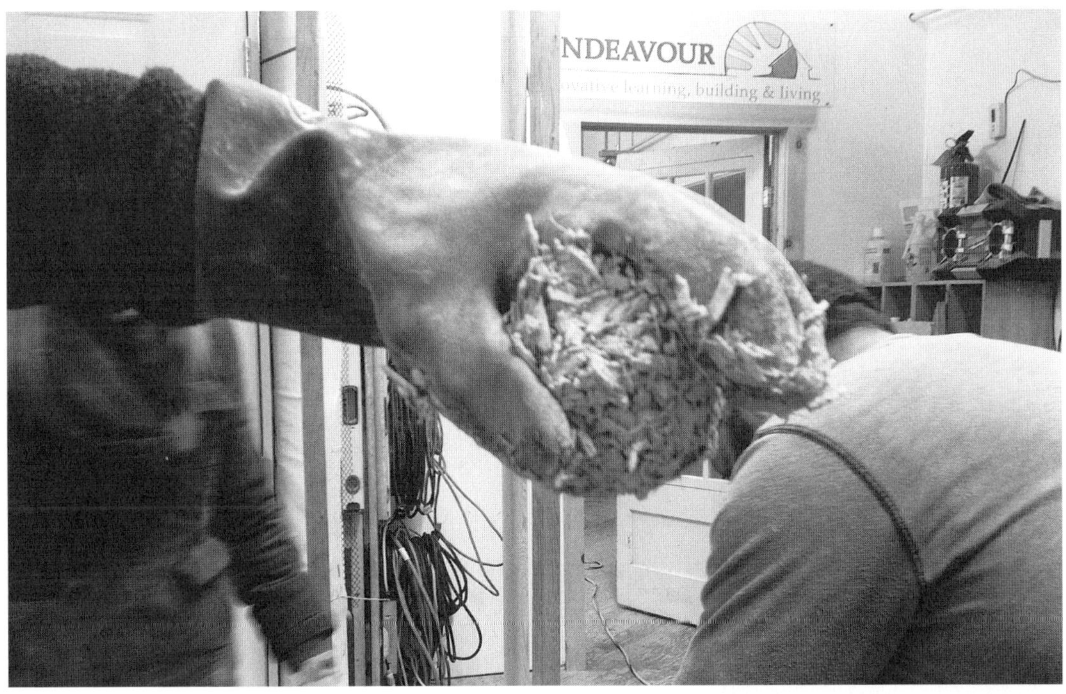

This ball of hempcrete is perfect for light- to medium-density mixes. It has been hand formed into a coherent ball and can be held from the top without breaking apart. It is fragile enough that a light squeeze will cause the ball to fall apart.

its shape but if shaken or disturbed it will break apart.

For medium density mixes, the squeezed ball should maintain its integrity if held from the top and given a light shake.

For high density mixes, the squeezed ball should maintain its integrity even if handled roughly.

If the ball drips water when squeezed, there is too much water.

When experimenting with water quantity, it is best to start with a mix that is too dry and slowly add water until the right consistency is achieved. Once a mixture has become too wet, adding more hemp and binder will not cure the problem, as the hemp in the wet mix will very quickly absorb the excess water and it won't be freely available to the added dry ingredients.

Visually, hempcrete mixes look like crumbly oatmeal. Upon being dumped out of the mixer it should not be a wet mass and should have no cohesion unless squeezed or tamped.

Hempcrete is loose and crumbly until tamped.

Mixing Procedure

Once the mix proportions have been established, it is helpful to translate the weight quantities to volume quantities that can be easily used by the person doing the mixing. Ideally, the volume quantities will equate to units that serve the mixer well, such as full bags of binder or full buckets, so that mixing can be done quickly and efficiently.

Buckets are marked so that the person making the mix can easily measure ingredients by volume and add them to the mixer.

Typical horizontal shaft mortar mixers do an adequate job of mixing hempcrete, and are widely available in a range of sizes.

Best results are achieved when the hemp hurd and the binder ingredients are dry mixed together until the binder is well distributed over the hemp. Water is then added evenly as mixing continues. If water is dumped into the mix too quickly it will be concentrated in one area of the hemp, and it will be absorbed by the hemp in that part of the mixer, robbing the rest of the mixture of important hydration. Adding a slow, steady stream of water and moving the stream back and forth across the mixer will allow for even distribution.

Allow mixing to continue for 1 to 5 minutes, until the whole mix is evenly moist. Stop the mixer and perform a quick squeeze test before emptying, and add water if required.

Mixing Options

The high volume of hempcrete required for most insulation projects favors the use of mechanical mixing machinery. There are two common options for large-batch mixing of hempcrete:

Horizontal shaft mortar mixer: This type of machine is widely available in North America, and, while it is designed to mix masonry mortar, it will do a sufficient job of mixing hempcrete. Portable mortar mixers range in volume from 4 to 14 ft^3 (0.11 to 0.4 m^3). Larger-capacity mortar mixers will be more efficient for mixing hempcrete. Mortar mixers can come with gas- or electric-powered motors, and can be rented from most tool rental centers.

Vertical shaft mixer: This breed of mixer is less common in North America (it is a European design), but does a better job of mixing hempcrete. Portable versions range in volume from 4 to 12 ft^3 (0.11 to 0.34 m^3). Gas- or electric-powered versions are available, but rentals are much less commonly available than horizontal shaft machines.

Don't use a *barrel-style* cement mixer. This is the most widely available type of mixer, but it is entirely inadequate for mixing hempcrete. The rotating barrel is designed to lift heavy gravel aggregate, which then falls and mixes as it tumbles. The lightweight hemp will become glued to the sides of the barrel as soon as water is introduced and it will be an exercise in frustration and failure.

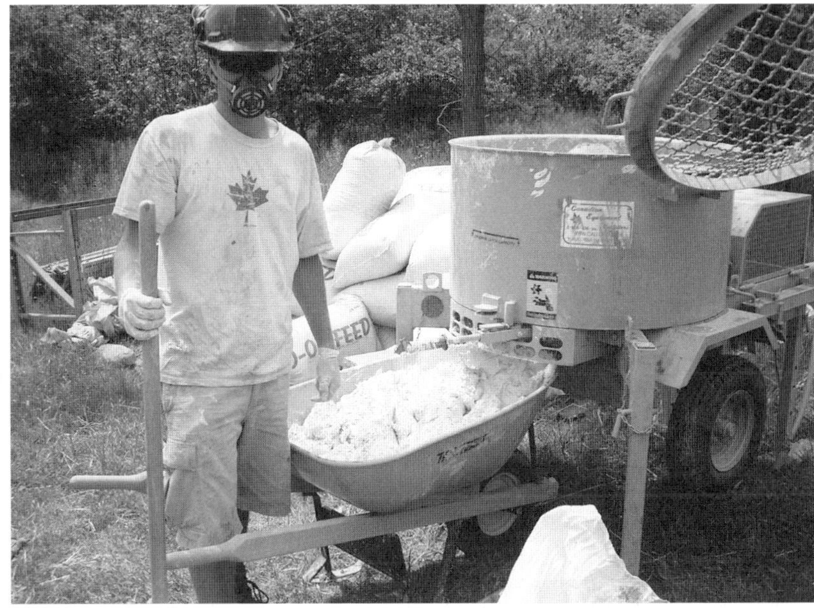

Specialty vertical shaft mixers are very good at mixing hempcrete, but may be harder to source than horizontal shaft versions.

Small-batch mixing

Smaller batches of hempcrete can be mixed in barrels or wheelbarrows with a drill mixer, shovel or hoe. The mixer will not be able to keep up with a crew attempting to fill large areas with insulation, but this method can be suitable for smaller projects or smaller volumes, such as window insulation and trim.

This crew is properly attired for mixing hempcrete, but they won't get viable hempcrete out of this kind of barrel-style concrete mixer.

More labor-intensive mixing methods are viable for small volume projects.

Chapter 9

Construction Procedure

Framing

THE FRAMING OF WALLS, floors and roofs for the use of hempcrete follow the same procedure as typical wood framing. Excellent resources are available to guide builders through the process of building code-compliant framing (see Resources).

For floor and roof framing, there will be no difference in procedure when using hempcrete insulation, assuming the framing dimensions have been specified to handle the dead load of the hempcrete insulation.

Procedures for framed walls will vary based on the style of frame wall being used (see Design Options chapter). Frame walls are typically laid out and assembled on the floor deck of the building. In conventional construction, structural sheathing is often affixed to the wall frames while they are lying on the floor as a time saving measure, since it is easier to place large, heavy sheets of sheathing in this orientation, rather than having to lift, hold and fasten them vertically once the wall has been raised. The installed sheathing also keeps the wall frame square while it's being raised; eliminating the process of squaring the wall after it has been raised. Cutting out window and door openings in the sheathing is also easier while the wall is horizontal.

Whenever permanent sheathing is used on a hempcrete wall, it is best to install it while the wall is horizontal. In some cases, it may be faster and easier to install temporary formwork in this manner as well.

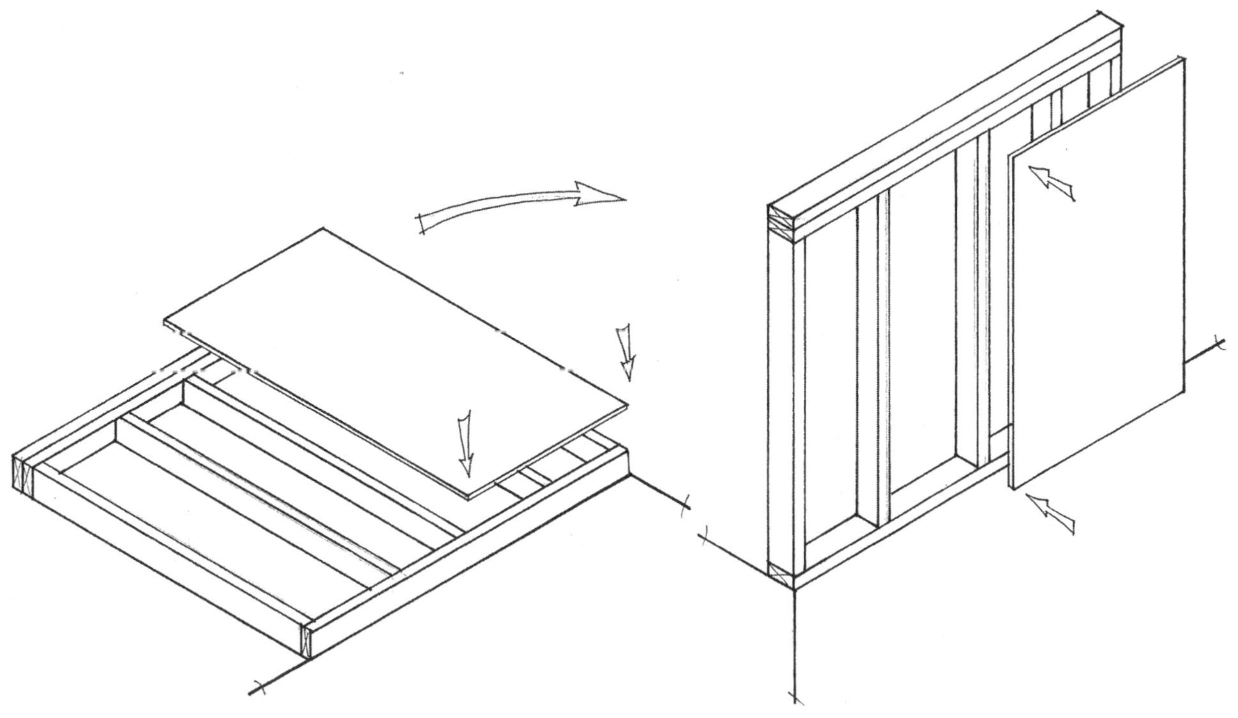

Formwork/Shuttering

The need for temporary and/or permanent formwork is the real difference in process between conventional framing and hempcrete walls. The installation and removal of temporary forms is a step unique to hempcrete (and other packed/tamped insulation materials), meaning that the time spent installing and removing temporary forms adds labor input and cost to building hempcrete insulation. This makes it important to develop a streamlined formwork system to minimize additional time and cost.

Permanent, permeable sheathing makes an attractive option for at least one side of a hempcrete wall because it removes the additional time and cost of applying and then removing formwork from one side of the wall.

Temporary formwork is used when the wall will have a plaster finish and/or to allow the hempcrete to dry out before a permanent sheathing is applied. Plywood or oriented strand board (OSB) are the most common materials to use for temporary formwork but should never be used as permanent sheathing as the low permeability inhibits drying.

Slip forms on at least one side are required so that the hempcrete can be placed and tamped by the installers. Plywood or OSB are typical forms.

A) One side fully shuttered, one side with slip form

By forming one whole side of the wall, the formwork can be added to the wall while it is lying flat, saving on time and effort. For temporary formwork, the quantity of material needed for forming will be much higher. If the forming material can be re-used on another element of the building or on another hempcrete job, this may be worthwhile.

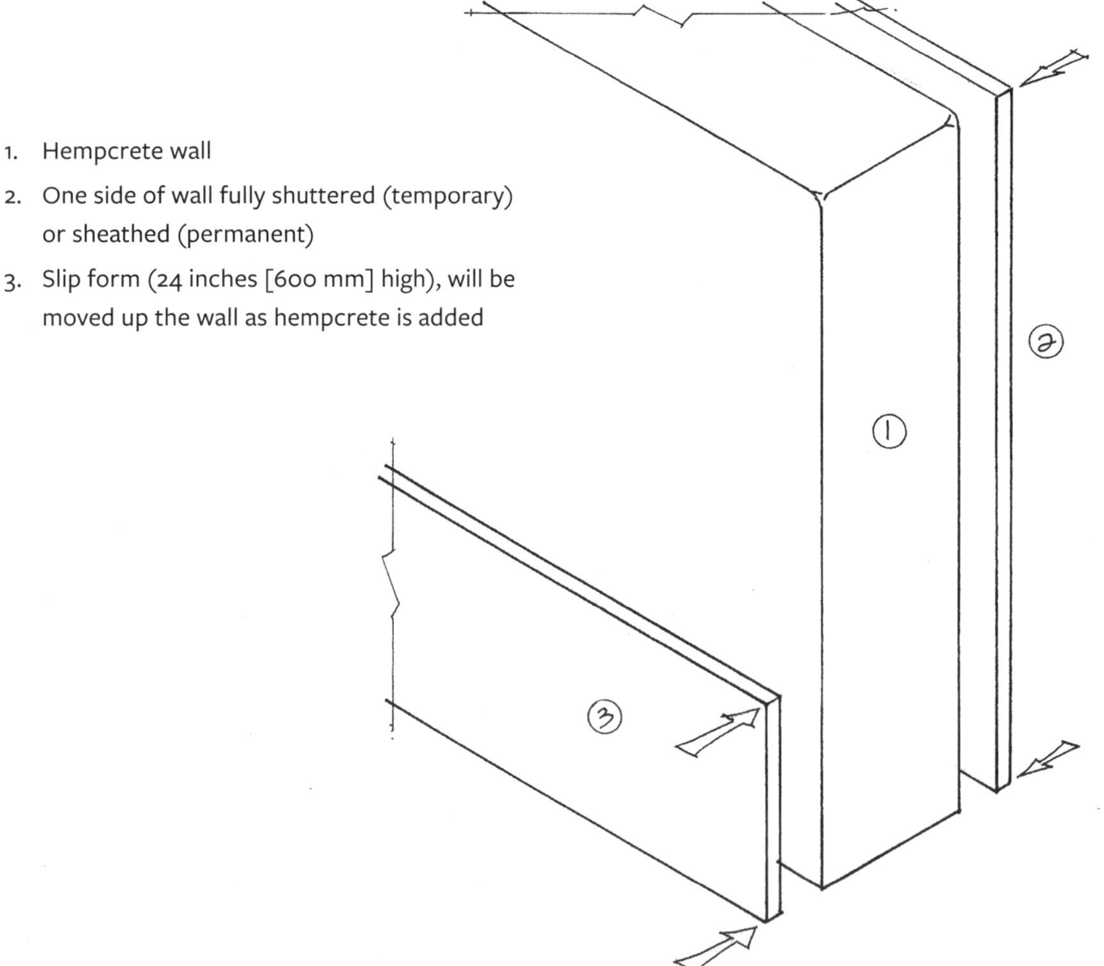

1. Hempcrete wall
2. One side of wall fully shuttered (temporary) or sheathed (permanent)
3. Slip form (24 inches [600 mm] high), will be moved up the wall as hempcrete is added

B) Slip forms on both sides

Slip forms require less material, but more labor to prepare, install and remove.

1. Hempcrete wall
2. Exterior slip form (24 inches [600 mm] high), will be moved up the wall as hempcrete is added
3. Interior slip form (24 inches [600 mm] high), will be moved up the wall as hempcrete is added

C) Formwork on spacers

The use of spacers is only required for wall options where the final plane of the hempcrete insulation extends beyond the framing on one or both sides of the wall. The spacers can be cut from common ABS or other rigid plastic pipe. Installation of this type of formwork takes a bit longer than when attached directly to framing.

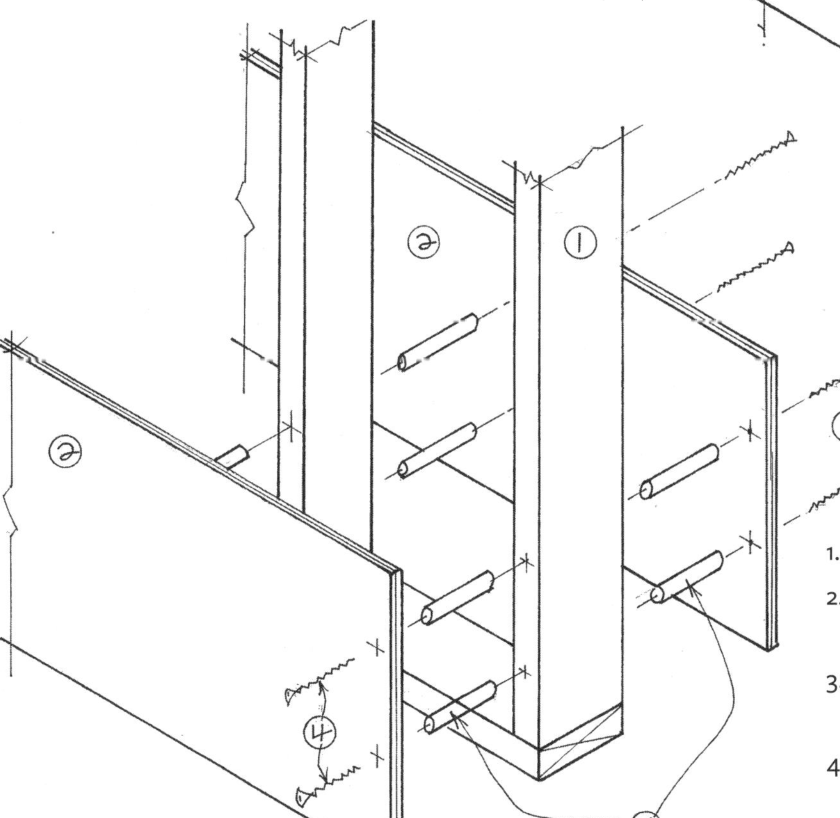

1. Centered frame
2. Slip forms (24 inches [600 mm] high), will be moved up the wall as hempcrete is added
3. Plastic pipe spacers, cut to desired length to achieve intended depth of hempcrete insulation
4. Long screws to extend through plastic pipe spacer and attach firmly to framing

Fasteners for temporary formwork should have pan heads or integral washers, not tapered heads, to prevent the screws from driving into the forms, ruining the forms and making it difficult to remove the buried screws. Choose screws with positive driver bit engagement (Robertson or Torx) or a hexagonal head.

Minimalist formwork is adequate for hempcrete walls. The material is light and does not exert much pressure on forms. Those used to forming for concrete will be surprised that a simple sheet of ½-inch (12.5 mm) plywood or OSB without any backers or reinforcement will be adequate in most situations.

Timing for formwork will depend on the size and efficiency of the hempcrete mixing and installation crew. If a large and/or experienced crew is involved, one complete lift of formwork around the building should be completed prior to the start of mixing.

Interior or exterior loading of hempcrete should be decided prior to installation of formwork. This decision will be guided by factors like wall height, site access, availability of scaffolding, number of stories and order of site operations. If using permanent, permeable exterior sheathing, then slip forms will be on the interior.

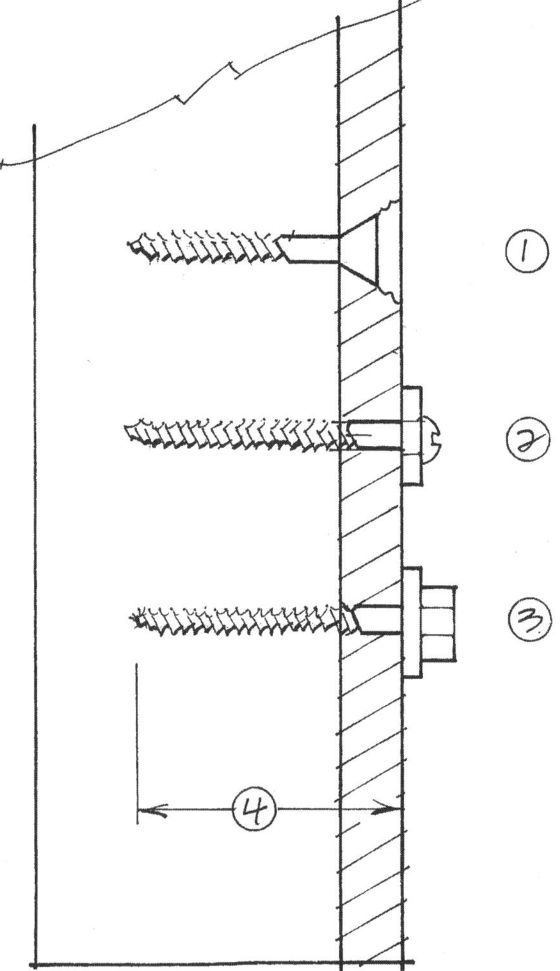

1. Tapered head screw. These will inevitably be driven too deeply into the forms and will be difficult to find and remove, especially once the forms have some hempcrete on them.
2. Pan head screw with Robertson or Torx head. These will stay on the surface of the form and be easy to locate. The heads may fill up with hempcrete over time.
3. Hexagonal head screw. These will stay on the surface and will not be affected by hempcrete.
4. Choose a screw length that is 2.5 times the thickness of the form material (½ inch [12.5 mm] forms use 1¼ inch [32 mm] screws). Screws that are too long will waste time every time they are driven and removed.

Steps

Once the framing has been completed and the type of formwork has been chosen, the process of installing the hempcrete is ready to begin.

1. ***Install as much formwork as is practical prior to mixing hempcrete.***

   It is best to take care of trickier areas like window openings before starting to install hempcrete, as custom cutting and fitting of forms in these places takes time that could slow down the installation.

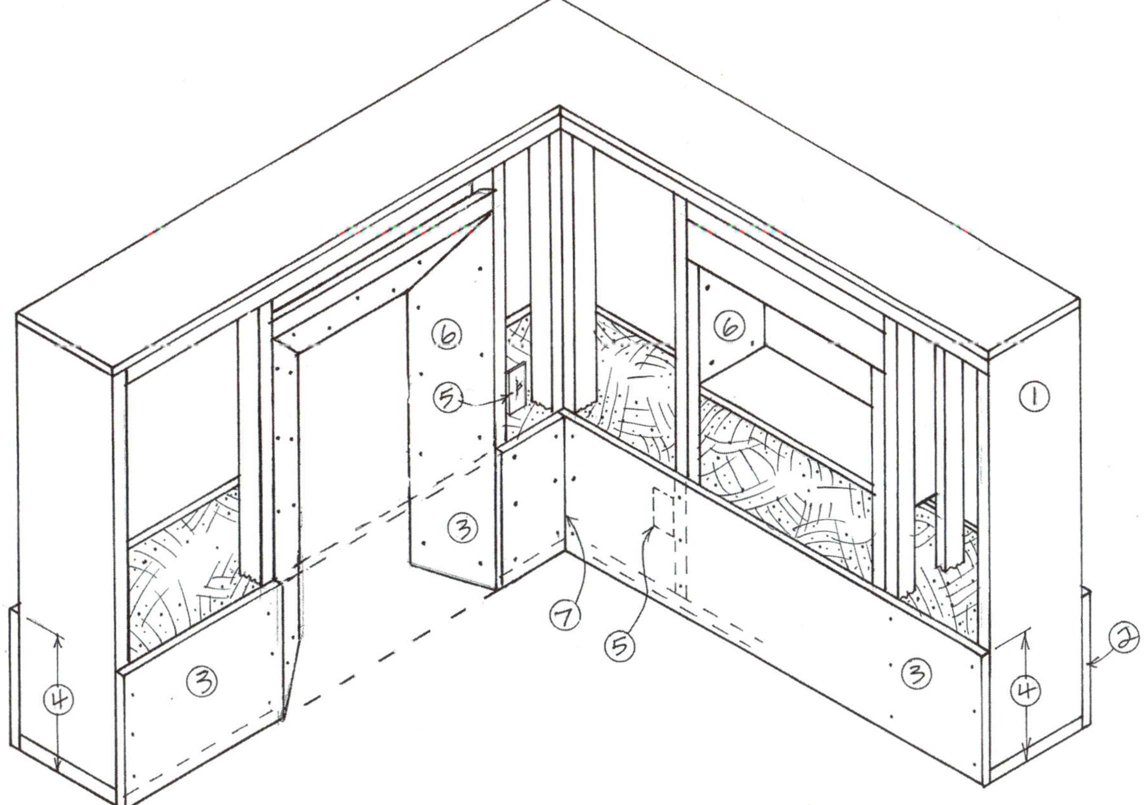

1. Wall, framed as per plans.
2. Exterior form. Can be full height of wall or slip form. One complete round of formwork should be fastened before hempcrete mixing begins.
3. Interior slip form. One complete round of formwork should be fastened before hempcrete mixing begins. Run form boards across openings to minimize cutting.
4. Height for slip forms is typically 24 inches (600 mm). This can be achieved by ripping a typical 4 foot wide sheet material in half, and is a comfortable reach for installers and tampers.
5. Any electrical boxes or other services and penetrations in the wall will need to be cut out of the form boards. This can be a time consuming process.
6. Formwork should be placed around the top and sides of all door and window framing. Leave the sill open to facilitate loading hempcrete from above. Flared, angled or curved openings will take extra time to form. Avoid having this formwork protrude beyond the face of the wall to avoid interfering with slip forms when they are placed in these areas.
7. Interior and exterior corners require secure attachment for forms, especially when centered framing and spacers are being used.

Shovels work well for placing hempcrete up to a comfortable height, and then small buckets become more practical.

2. *Begin mixing hempcrete.*

The mixing station should be well positioned on the site for easy access to all the materials and to the source for water; clear paths for the movement of wheelbarrows or other means of delivery should be established and kept free of obstacles. The mixer and all those working close to the mixing station should wear adequate breathing and eye protection, waterproof gloves, and long sleeves, as binder ingredients are caustic to skin and lungs.

1. Start mixer.
2. Add appropriate volume of hemp hurd.
3. Add appropriate volume of binder.
4. Mix until all hemp hurd is well coated in binder.
5. Slowly add water to the mix until the desired consistency is achieved.
6. Empty mixer into wheelbarrow or other means of delivery.
7. Repeat!

A mix of hempcrete will be workable for 30–90 minutes, depending on the strength of the hydraulic reaction of the binder. The mixer should work to keep the installers readily supplied with hempcrete without getting too far ahead. Keep finished mixes covered and out of direct sunlight. If a mix begins to set before it is installed it can be spread out on the ground and allowed to dry, and the dried material can be added as a small portion into new mixes until it is used up.

3. *Place hempcrete into forms in shallow lifts.*

Small buckets or shovels can be used to move hempcrete into forms. Spread the material out in the forms and build it up to a depth of 4 to 8 inches (100 to 200 mm).

4. *Press hempcrete into place.*

A gloved hand can be used to press hempcrete, or pressure can be applied using a piece of wood that is comfortable to grip and long enough to reach the bottom of the form. The word "tamp" implies a force that is greater than what is required. Your team should have some experience pressing hempcrete into sample forms, and be familiar with the feel of the mix as it is reaches the correct consistency.

The pressing should ensure that hempcrete fills all voids, paying particular attention to angles, corners and obstructions like electrical boxes. We tend to put a bit more pressure on the hempcrete where it contacts temporary forms to consolidate the surface that will be exposed for plastering. This step is unnecessary for permanent sheathing.

Remember that over-compressing the hempcrete will result in less thermal performance and will use more material than calculated. Quality control at this stage cannot be over-stressed.

A tamping stick with a relatively wide contact area can prevent applying excessive pressure. A gloved hand may be the best option for pressing hempcrete into forms.

5. *Raise slip forms when full.*

Typically, the forms for light- and medium-density hempcrete can be moved immediately, and the exposed hempcrete will maintain its form.

Remove the screws from the form boards, starting at the far edges and working toward the middle. This will prevent the form from dropping quickly when the last screw is removed. To prevent pulling hempcrete away from the wall as the form is removed, slide the form horizontally and/or downward before pulling it away from the wall surface.

The form can be refastened higher up the wall. Provide about 4 to 6 inches (100 to 150 mm) of overlap from the bottom of the form board to the top of the hempcrete insulation. If this is not done, the pressing of the next layer of hempcrete can cause the material around the bottom edge of the form board to bulge outward.

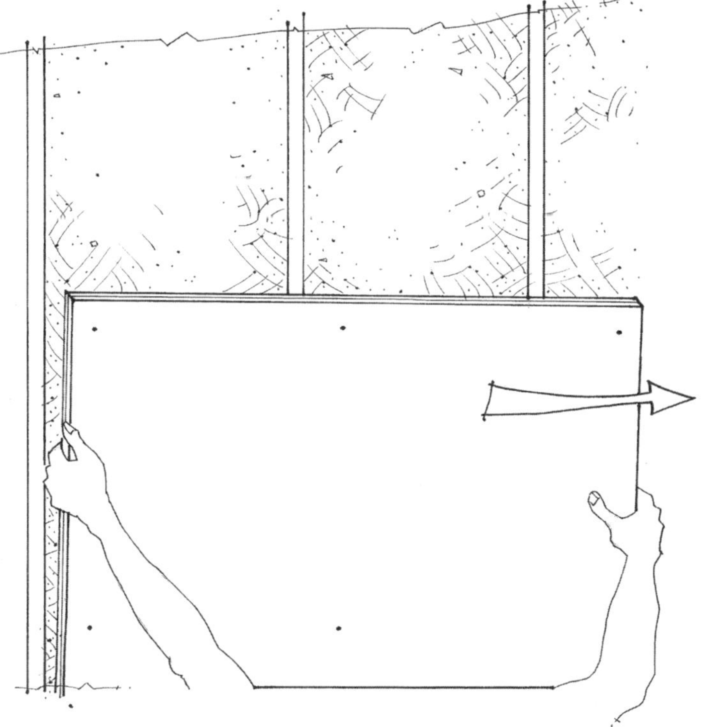

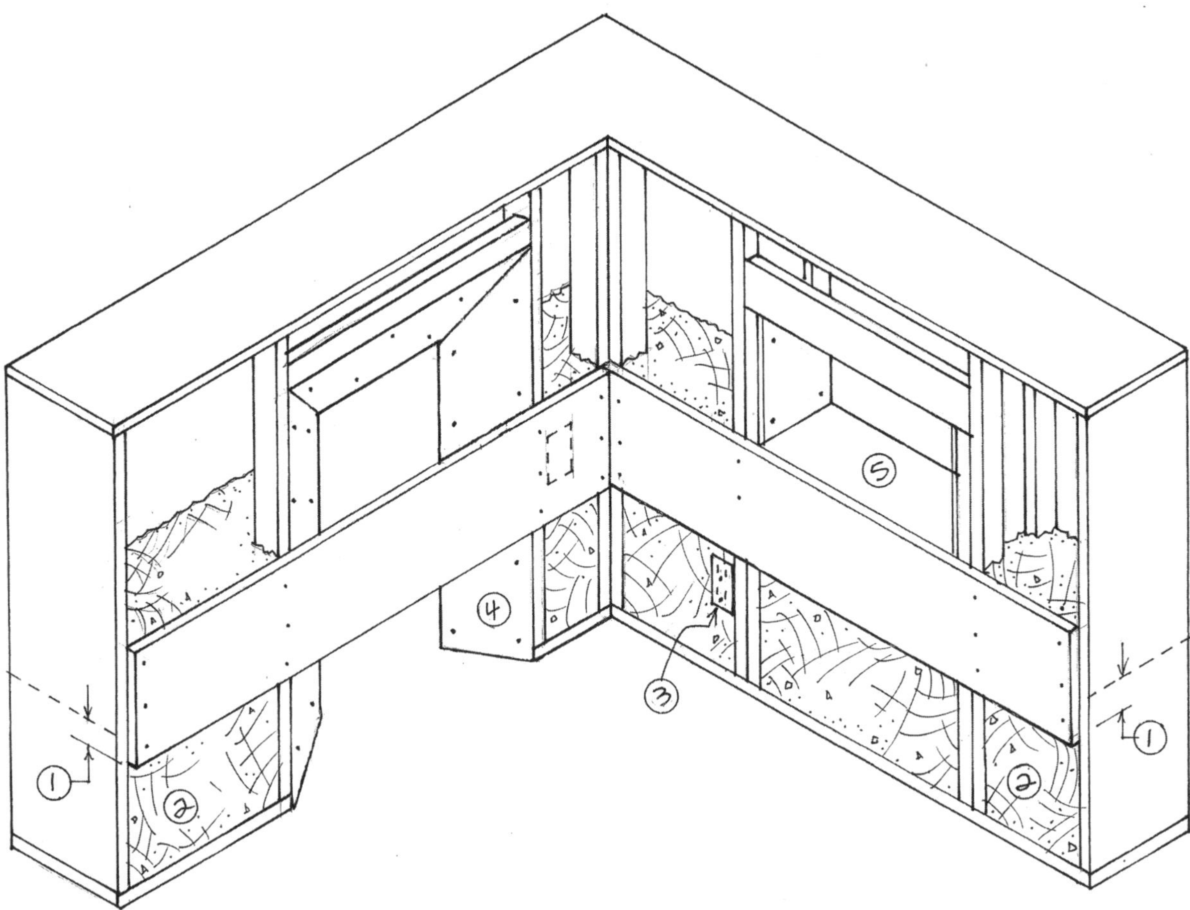

1. Overlap form board with top of hempcrete by 4 to 6 inches (100 to 150 mm).
2. Hempcrete insulation should be well consolidated on the surface, and maintain its shape and integrity. If this is not the case, halt work and troubleshoot.
3. Inspect carefully around corners, edges and electrical boxes to ensure that hempcrete has filled all voids and presents an even surface. If any areas have been missed, press some fresh hempcrete into the area to fill the void and ensure a smooth finished plane.
4. Leave door and window framing in place until all hempcrete is placed, and allow form boards to cover openings to minimize the amount the forms need to be cut.
5. Windowsills will be filled from the top. A form can be screwed on top once filling is complete to level the surface.

6. *Slip forms do not go all the way to the top on one side.*

Hempcrete can be placed from the top until it is no longer practical to install from that direction. With the form all the way to the top on one side, hempcrete is installed from the front and pressed horizontally to fill the top of the cavity.

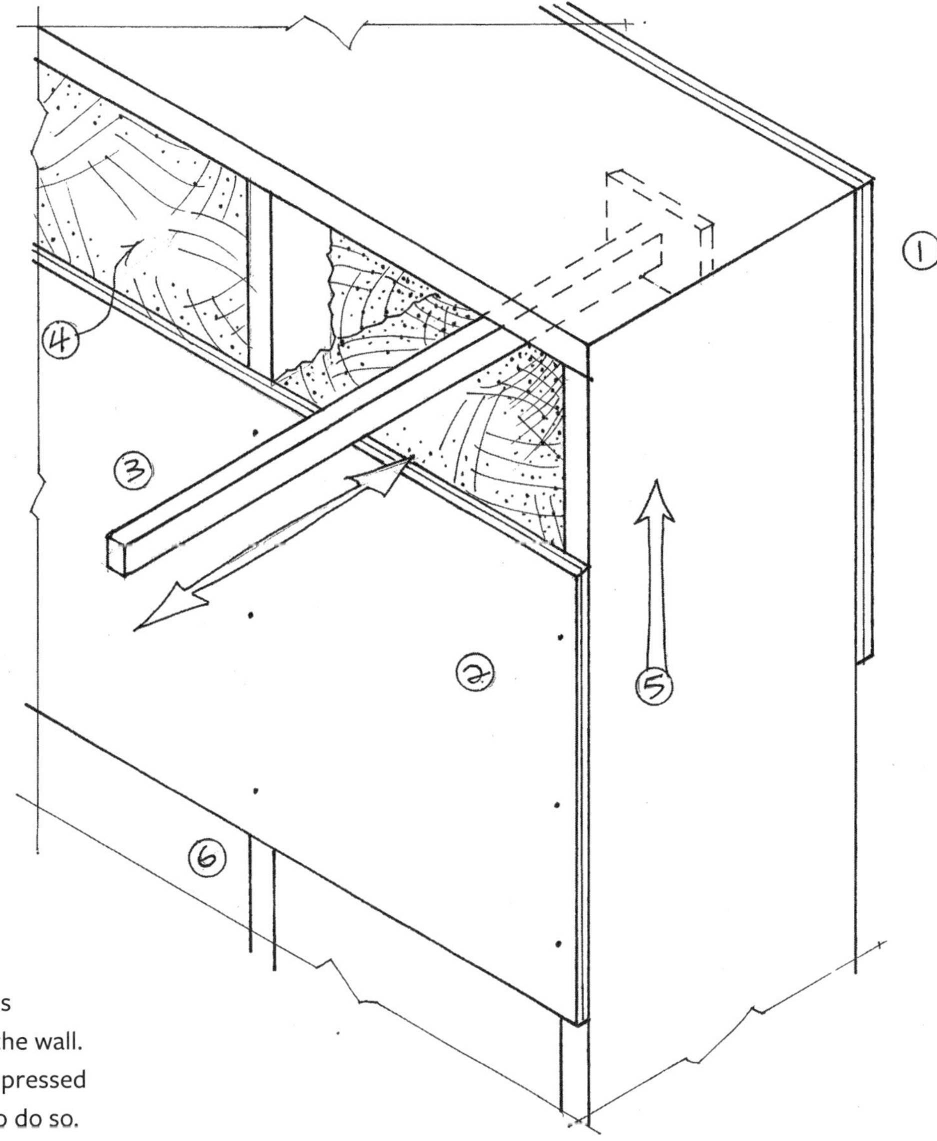

1. Form board extends to top of wall on one side.
2. Opposite form board is left 8 to 10 inches (200 to 250 mm) down from the top of the wall. Hempcrete is loaded from the front and pressed downward until it is no longer possible to do so.
3. Using a tamper or a gloved hand, press hempcrete against back form and build the insulation toward the opening, being sure to match the level of pressure used elsewhere in the wall.
4. Press the hempcrete into the opening until it bulges slightly beyond the face of the form board.
5. When the entire opening has been filled slightly beyond capacity, remove the form and screw it into place at the very top of the wall. This will compress the hempcrete that was proud of the frame and provide a flat surface to the top edge of the wall.

As soon as the wall is finished, inspect for voids and fill them with fresh mix.

A finished hempcrete wall is satisfying to look at!

7. ***Remove all formwork and inspect the wall for voids.***

   If using spacers, they can be removed while the hempcrete is fresh by using pliers and turning them back and forth before pulling them out. Spacer holes and any areas that were missed when filling forms — most commonly at corners and around electrical boxes — can be filled with fresh mix and a tamper used to press the surface flat.

   Ridges or other unevenness in the wall can be carefully brushed or shaved back using a plaster rasp. Sight along the wall and be sure that a flat, even plane has been created.

Estimating Installation Labor Time

Once the walls are framed and initial forms are set, a crew consisting of one mixer, one "runner" delivering mix, and two to three installers (who also move forms as required) can easily place 25 to 40 cubic feet (0.7 to 1.15 m^3) of material per hour.

For walls that are 8 feet (2.4 m) tall and 12 inches (300 mm) wide, this translates to 22 to 35 lineal feet (6.7 to 10.6 m) per day for a crew of four to five people.

Curing and Drying Time for Hempcrete

As a thick and moist insulation, hempcrete requires a period for curing and drying that is not typical of most other insulation types.

The *curing* of hempcrete happens in two phases. First, the hydraulic setting action of the binder is activated by water, and the resulting chemical set of the binder begins immediately. A high percentage of the chemical set occurs within the first 24–48 hours. All the structural properties required of the hempcrete are typically met within this short window of time, and, though the binder will continue its hydraulic set for days or weeks, the degree to which the chemical set is completed is not so important to the functionality. Secondly, the hydrated lime begins the process of re-carbonizing, absorbing atmospheric CO_2. It gains in strength as this happens, and the process can take years or even decades to approach completion. Again, the strength gained in the process is not important to the functional properties of the hempcrete.

The *drying* of hempcrete is the critical element, as the water added to the mix that has been absorbed by the hemp must be released. Due to the porosity of hemp hurds and the thickness and density of the material, the moisture levels in hempcrete can take quite a while to equalize with the surrounding atmosphere. Several factors will affect drying times:

- Moisture content of the hempcrete mix when placed in the walls. This is the largest determining factor, and is dependent upon the mix ratio and the effectiveness of the mixer in distributing the water content to the binder before it can soak into the hemp hurds. Using as little water as possible to make a viable mix will help to minimize drying time.
- Relative humidity in the atmosphere. Dry air will draw moisture out of the walls quite quickly, but high relative humidity can arrest the drying process completely. In very humid conditions, the use of industrial dehumidification may be necessary.
- Airflow around the hempcrete. If moving air (especially dry air) has contact with the hempcrete, drying will be faster. Having both sides of the wall exposed to the air will promote drying faster than if one side is behind permeable sheathing.

Time for this drying to take place must be built into the construction schedule. Rushing and covering hempcrete that is still too damp can at least affect thermal performance and at worst cause moisture-related problems such as mold or rot. While the hempcrete itself is quite resistant to mold and rot due to the presence of the lime binder, moist hempcrete insulation can raise the moisture level of wood framing and sheathing to potentially problematic levels.

A wood moisture equivalent (WME) probe meter can be used to ensure the walls are sufficiently dry prior to covering. Ideally, the center of the hempcrete should be in the range of 12–16% moisture content (MC) before being completely covered. However, a permeable sheathing or plaster can be applied safely at anything up to 20% MC. Some builders will cover up at 25%, but anything above that may be problematic. There has been no testing of hempcrete closed in at different moisture levels, so it is best to be conservative and ensure that the insulation is as dry as possible.

Hempcrete walls can dry sufficiently in as little as two weeks, and can take as long as eight weeks. Electric fans and dehumidifiers can speed up the drying process if necessary.

Plaster finishes can be discolored if applied over hempcrete that is too wet, because tannins in the hemp hurd are transported through the plaster as the moisture leaves the insulation, and these are deposited in the plaster. This is an annoyance more than a building issue, but can be avoided by ensuring hempcrete is dry.

Chapter 10

Finishes

THE TYPES OF FINISHES that can be applied to hempcrete walls fall into two categories:

- Plaster, including lime-based plasters for exterior use and lime-, clay- or gypsum-based plasters for interior use.
- Standard sheet finishes on the interior, including drywall and MgO board and structural, permeable sheathing on the exterior, including wood fiber board, MgO board or exterior gypsum sheathing, plus any type of rain screen siding, including wood, metal, composite boards, brick or stone.

All exposed wood to be covered in plaster must be meshed or treated with a glue-sand mixture to provide mechanical grip for the plaster.

Plaster Finish

Surface preparation

Hempcrete insulation makes an excellent substrate for plaster, providing a great deal of mechanical "key" in the form of small voids between pieces of hemp hurd, a flat surface that can be covered without variation in plaster thickness (a major cause of cracking) and for lime-based plasters the lime binder in the hempcrete will create a strong bond between the two layers.

Any exposed wood must be treated so that it offers the plaster a similar degree of mechanical key as the hempcrete. This can be achieved by stapling mesh (fiberglass, plastic or metal) to the wood and allowing the mesh to carry over onto hempcrete by one or two inches (25 to 50 mm). Wood can also be coated in a glue and sand mixture (gluten- or casein-based glue mixed with sharp sand) to help the lime plaster adhere.

Careful consideration of all flashing and trim details must be undertaken in order to prepare for plastering. Plaster stops may need to be installed around the edges of the walls and where plaster meets windows, doors and trim.

Plaster mixes

Plaster is installed in two or three coats, depending on the desired thickness and finish, the skill level of the plasterer and the quality of the edge details.

The first and second coats are typically ¼ to ⅜ inch (6 to 9.5 mm) thick. The surface of the plaster should be scratched or roughed to provide mechanical key for the subsequent coat. The third coat is typically ⅛ to ¼ inch (3 to 6 mm) thick.

Lime wash or silicate dispersion paint can provide a suitably permeable color coat.

Exterior plasters are generally always lime-based. Typical recipes for the first and second coats include:

- 1 part hydrated lime to 2.5 to 3 parts stucco sand. Hydrated lime plaster achieves its strength by re-carbonizing with atmospheric CO_2, and so the setting process is relatively slow. It is common to leave each coat of this plaster at least one to two weeks before applying the next coat, and to keep the plaster from exposure to direct sunlight and strong wind. The plaster is misted with water several times over the first several days to keep it hydrated.
- 1 part hydraulic lime to 2.5 to 3 parts stucco sand. Hydraulic lime plaster achieves a portion of its strength from a chemical set in the presence of water, and after this it re-carbonizes slowly. The chemical set allows for the application of the next coat within 24 to 72 hours. Protection from direct sunlight and strong wind, as well as misting with water is also necessary for hydraulic lime plaster.
- 1 part hydrated lime, 1 part hydraulic lime, 5 to 6 parts stucco sand. This hybrid plaster achieves the early setting of the hydraulic lime plaster but offsets the higher cost of the hydraulic lime.
- 1 part hydrated lime, 1 part metakaolin, 5 to 6 parts sand. This plaster has a chemical set similar to the hydraulic lime plaster but can have a lower cost.
- 1 part hydrated lime, 0.5 part Portland cement, 3.75 to 4.5 parts stucco sand. This plaster loses some of the permeability of a lime plaster, but has a 24-hour set time and requires less tending.

All of the above lime plasters can benefit from 0.25 to 1 part of chopped hemp fiber to provide tensile strength. These recipes are also suitable for interior plastering.

Clay plaster can be used over hempcrete on the interior walls:

- 1 part clay, 2.5 to 3.5 parts stucco sand, 0.5 to 1 part chopped hemp fiber. This plaster can be used as a base coat and second coat.
- 10 parts clay, 4 parts finely sifted sand, 1 part calcium carbonate, 1 part flour paste, 0.5 to 1 part pigment (if desired). This plaster makes a good interior finish coat.

Plaster application

All of the above mixes are combined with water in a mortar mixer to form a workable, plastic consistency, and then troweled onto the wall.

Plastering is an interesting finish, one that can take years to master, and yet is also inviting to beginners.

There are many excellent resources for learning about plastering, and it is recommended that a new plasterer spend some time becoming familiar with the mixes and applications before undertaking a hempcrete plastering job.

A lime wash adds color and protection to an exterior lime plaster.

Wallboard Finishes

The interior of a hempcrete wall can be finished with conventional wallboard treatments, with the boards screwed to the wall framing and the seams taped, mudded and sanded to provide the desired finish. Drywall and MgO board are the most typical wallboards, although clay board is gaining popularity in Europe.

The installation and finishing of wallboards is a common trade, and information regarding the use of these products is widely available.

Conventional drywall can be applied over unconventional natural fiber insulation. In this case, the final finish is lime paint, which helps maintain the look and feel of natural materials.

1. Wall core
2. Vertical strapping affixed to wall, 3/8 to 3/4 inch (9.5 to 19 mm) typical
3. Siding material
4. Top of ventilation channels exhaust upwards to soffit
5. Top of ventilation channels exhaust outwards
6. Bottom of ventilation channels protected by insect and rodent-resistant screen/mesh

Rain Screen Exterior Finishes

The exterior of a hempcrete wall can be finished with a wide variety of siding. Each of these is applied to create a ventilated rain screen. This type of siding installation provides a great deal of resilience for a building by adding a durable and weather-resistant siding over an air space that both prevents water from reaching the wall core and provides an opportunity for drying if the wall core experiences high humidity or wetting. The lifespan of the siding is improved by the presence of the air space behind, and when the siding eventually needs replacing, the job does not affect the integrity of the wall core.

Board or *sheet-style siding* comes in many forms, including vertical and horizontal varieties of wood siding, wood shingles, metal sheets and shingles, and composite planks and panels. All of these are fastened to the wall using vertical or vertical and horizontal wood strapping, with the vertical strapping creating the air channel between the face of the wall core and the back side of the siding.

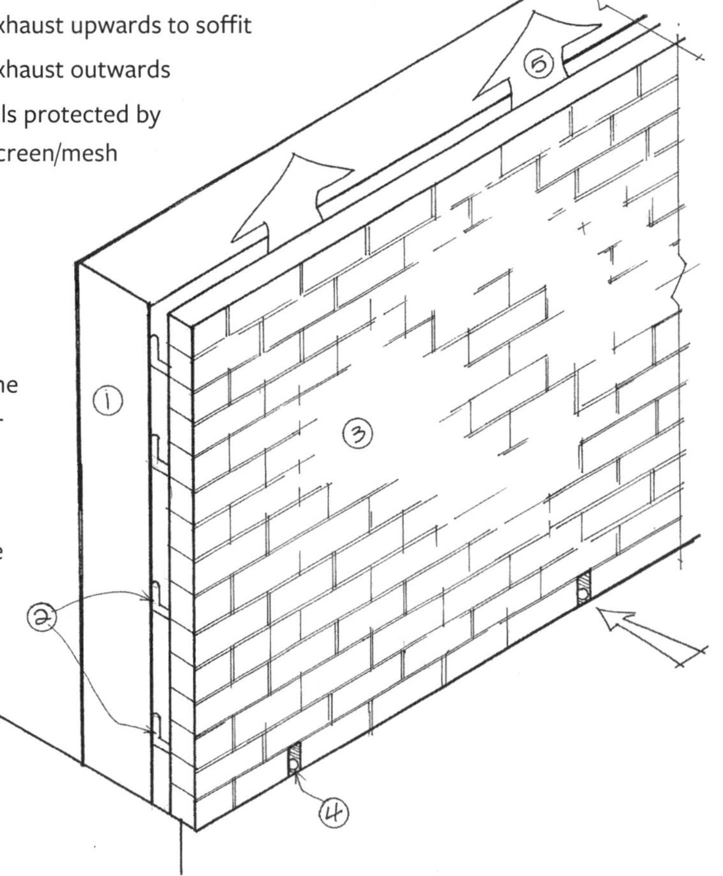

1. Wall core
2. Metal masonry ties fixed to the wall and embedded in mortar joints at specified intervals
3. Masonry siding
4. Weep holes in bottom course
5. Ventilation to soffit

Masonry siding includes brick and natural and manufactured stone. These materials are installed over a hempcrete wall in a manner that is identical to other conventional installations.

Many factors go into deciding on siding for a building, including cost, aesthetics, environmental impacts and durability. The book *Making Better Buildings* offers excellent guidelines for making appropriate choices.

Chapter 11

Maintenance and Renovations

HEMPCRETE INSULATION is durable and resilient. There are no required maintenance intervals for the material, and it should last the lifetime of the building.

As the insulation is part of a conventional frame wall system, renovations or alterations to the building will meet the expectations of any experienced tradesperson. Though the hempcrete may not be as simple to remove from cavities as loose fill or batt insulation, neither is it difficult to do so. Hempcrete cuts very easily with a handsaw or a reciprocal saw and the material can be broken up and added to a new hempcrete mix when the work is complete and it is time to insulate again.

Chapter 12

Building Codes and Permits

THE USE OF HEMPCRETE INSULATION in North America is now into its second decade. While the number of hempcrete-insulated buildings is still relatively small, there is a reasonable precedent on this continent in a variety of jurisdictions and climate zones, and a stronger precedent and more examples in the UK and Europe. However, there is no direct recognition of hempcrete insulation in any North American building codes, nor are there published standards in the U.S. or Canada. Even overseas, there are no jurisdictions in which hempcrete is directly recognized in codes, though the UK, France and Australia have all granted "certificates of conformity" to proprietary hempcrete formulations. Any permit sought for a building using hempcrete insulation will need to be done using "alternative compliance" pathways, using the available research base and test results to argue for the equivalency of hempcrete insulation.

Just insulation

Despite the lack of code recognition, the hurdles to getting a permit for a building with hempcrete insulation should be relatively low. Because hempcrete is only an insulation material in an otherwise conventionally framed building, there are no thorny structural issues that need to be addressed in the permit application. The main concerns of code officials are likely to be fire resistance and thermal performance, and there are numerous studies and tests that address these concerns.

Adhering to local process

Before leaping into arguments with local code officials about hempcrete, it makes sense to learn about the local permitting process and understand a bit about your building code. A great number of permit problems arise from applicants simply not following the proper permitting process. Procedural issues far outweigh issues with a particular building component such as hempcrete insulation. Be sure to understand the complete process for obtaining a permit; often there are multiple stages (planning permission, site-specific permissions such as sewer/septic connections and property setbacks) that must be completed in a particular order. Most jurisdictions offer guidelines for applying for a building permit, and these should be followed carefully. Most building departments have to process a great number of applications and are often understaffed; in these circumstances, incomplete or improper applications do not receive favorable treatment.

Code issues

Both the IRC in the U.S. and the National Building Code in Canada are model codes; they have been adopted by state, provincial and/or local authorities in versions modified to suit local needs. Be sure to understand which code you will be working from, and use the most up-to-date version of that code document.

Building codes cover a wide range of issues and topics, and there are many areas in which a set of building plans may not conform to the code that have nothing at all to do with a particular insulation type, like hempcrete. In the author's experience as a consultant, the majority of permit denials for hempcrete or other projects proposing "alternative" materials have to do with issues that are unrelated to the specific material choices. Plans can have problems related to zoning issues (lot lines and setbacks, overall height, grading, parking allocation), space allocation (minimum room sizes, means of egress, staircases and railings, window size and placement) and services (well/water, sewer/septic, HVAC) that have nothing to do with chosen materials or assemblies, and these are common problems experienced by conventional and alternative proposals alike. Addressing them requires an understanding of the codes, but does not directly influence the use of hempcrete.

Despite plenty of anecdotal evidence to the contrary, there is no legal justification in any North American building code for denying a permit because of the use of hempcrete system. At the same time, I am unaware of any building department that will respond to a general inquiry about whether or not they will permit hempcrete insulation with a general answer of "yes." Permits are not granted or denied on the basis of a single material choice, but for meeting a complete set of requirements that demonstrate the viability and safety of the entire structure.

All codes include provisions for working productively with materials that are not directly recognized by code prescriptions, or materials being used in ways that are not directly prescribed in the code. In making a permit application, you must understand what aspects of your proposed building do and do not meet the prescriptions of the local code. You must also acknowledge that plan reviewers are concerned with specifics, not generalities; permits are granted or denied based on the exact details of the proposal, and if the details are missing, inadequate or in contravention of the code, a permit will not be granted even if the idea is generally feasible.

Alternative compliance applications

Every building code has a mechanism for consideration of nonconforming materials and approaches. If you have identified that some elements of your design cannot be supported via code prescriptions, it is incumbent on you to understand the exact procedures used in your code jurisdiction to handle alternative compliance. While the paperwork requirements may vary, all such alternative compliance pathways operate on the assumption that the applicant will provide proof that the alternative proposal meets or exceeds the provisions of the prescriptive code requirements. Any performance parameter (structural capacity, fire resistance, thermal performance, etc.) that exists for a wall in the prescriptive section of the code must be demonstrably met or exceeded by the proposed alternative. Each of these performance parameters must be fully supported and documented.

Several options exist for demonstrating that an alternative solution meets or exceeds code requirements:

- **Past performance** An applicant can typically cite prior examples of the same or similar approach used successfully in the jurisdiction. Be sure to have adequate documentation of past performance to ensure that the approach was similar to what you are proposing, and to be sure that it was indeed a successful approach. Past performance or case studies from other jurisdictions or other countries may not be viewed as conclusive evidence, especially if the climates are different. The quality of documentation will be examined carefully, and if it doesn't stand up, it may not be recognized as proof of equivalency.
- **Testing data** It is best if the tests are done to a code-recognized standard, such as ASTM, ANSI or CSA. If the tests are not performed

to the standard used by the code, be prepared to show how the testing varies and how the results may be interpreted to show equivalency. For hempcrete, almost all the tests have been performed to British or European standards, so you will need to show that these standards meet the intent of the local codes.
- **Professional seal** A licensed architect and/or engineer can provide code equivalency assurance to the building department by applying their seal to the drawings and so ensure that to the best of their professional ability the alternative approach meets the intent of the code.

In some cases, it may be that all of these approaches are employed on an alternative compliance application. In all cases, understand that it is entirely up to you as the applicant to provide the information and any supporting interpretation to the building department. For better or for worse, building departments are reactive, not proactive. They are under no obligation to assist you with your documentation or ensure that it is complete. They are only obliged to respond to what has been provided in the application.

Code consultants are professionals that may be hired to assist an applicant with understanding the code and all the parameters that need to be addressed in order to put forward a complete application.

Rejections and appeals

It is important to know that a permit cannot be denied for any reasons other than code infractions or incomplete submissions. Every building code prescribes the manner in which a denial is presented to the applicant. In most code jurisdictions the procedure for a permit refusal involves a written response explaining the code infractions that caused the permit to be denied. This is intended to give the applicant a full understanding of where the application was found to be lacking and provide a blueprint for resolving the issues in a resubmission. If all of the code issues are fully addressed in a subsequent submission, then a permit should be issued. In many cases, there can be several rounds of rejection and resubmission. While this may be frustrating and time consuming, getting a building permit can be compared to taking a test where you must score 100%. Building departments cannot let any infractions they detect slip through without being addressed, so it is best to consider the application to be a multi-step affair.

Forming and maintaining a good working relationship with the plan reviewer is very helpful. At best, the plan reviewer will be acting as an advocate and will be assisting you with understanding where the plans fall short of meeting the code and making suggestions regarding how the deficiencies can be corrected. At worst, they are obliged to make your mistakes known to you, and you will have to figure out how to correct them.

Should there be a disagreement about code compliance, every code jurisdiction has an established route for appeals. Often, this involves taking the dispute to the Chief Building Official. Should this fail to resolve the issue, there will be a higher regional, state or provincial authority that will hear appeals, and the pathway to accessing the appeal should be provided to you. Many appeal processes are quasi-judicial and involve a hearing where both the applicant and the building department put forth their arguments and a panel renders a decision.

Preparation and patience are invaluable

Any application to a building department involving an alternative compliance element should be made well in advance of needing the permit to allow the process to go through a few rounds of back-and-forth. Expecting or, worse, demanding a fast turnaround for an alternative compliance application is to invite frustration and delays.

Any applicant willing to put the time and effort into making a viable and complete initial submission and diligent enough to follow through with any requests for changes or more information can expect to be rewarded with a permit.

Chapter 13

Experiments and Future Developments

HEMPCRETE DEVELOPMENT is not limited to formed wall, roof and floor insulation. Work is ongoing around the world in adapting hempcrete to many other potential uses.

Precast blocks and panels

The need for hempcrete to be formed and cured makes it a prime candidate for various types of precast products. This type of production eliminates the need for on-site mixing and placing of hempcrete and allows the curing/drying process to happen in controlled conditions outside the construction schedule of the building. Delivery and installation of ready-made blocks or complete wall panels would put hempcrete construction on par with other prefabricated building products, and eliminate some of its major drawbacks.

Currently, there is limited production of blocks and full wall panels in the UK. Until there is a healthy hemp industry in North America, this type of production is unlikely to happen here, but the technology and ability to move quickly into block and wall casting certainly exists.

Hempcrete spraying

There is some use of industrial spraying equipment to apply hempcrete in the UK and France. Though no examples of this method of application have yet been used in North America, it is an approach worthy of exploration. Reports from companies involved in spraying, as well as some academic research on the subject, show that spraying hempcrete can reduce application times and drying periods and lower the density of the installed material.

Just Biofiber of Calgary, Alberta, is one of a number of companies working to create prefabricated hempcrete building elements. These types of products may make hempcrete building more accessible to builders.

Should the North American market for hempcrete begin to expand, spray applications will likely begin to play a role in the commercialization of hempcrete.

Hempcrete and cordwood building

Cordwood building has a long history in North America. The system uses lengths of split softwood logs in a matrix of mortar and insulation to form walls that have some excellent characteristics. Cordwood walls make use of "waste" wood resources that wouldn't otherwise have found use as building material and can provide a reasonably well-insulated wall system.

In typical cordwood construction, a cement- or clay-based mortar is used to support

There aren't many examples of combining hempcrete with cordwood construction, but experience points to many benefits.

the logs on the inner and outer edge of the wall, and the voids around the logs are filled with sawdust insulation. The mortar in the system is not insulative, and there are often gaps between the mortar and the wood caused by differential expansion and contraction. Both of these factors rob the walls of thermal performance.

Hempcrete can make an interesting mortar for cordwood walls, as it can be used through the entire thickness of the walls. This eliminates the need to build up two distinct mortar beds and then backfill with insulation. Hempcrete can provide the strength of the mortar and the insulation factor in a single, full-thickness application. Hempcrete also expands and contracts at a rate very similar to wood, so using it will reduce cracks.

Work would need to be done to refine this hybrid into its own system, but it shows a great deal of promise.

Decorative and sculptural possibilities

Hempcrete has a fascinating consistency when it is wet. It can be formed into any shape and immediately has a cohesiveness that enables it to keep its shape, especially when some hemp fiber has been introduced into the mix. The light weight allows the material to "defy" gravity and be built up quickly and with little or no support required. At the same time, it remains moist and workable for a long time. It can also be cut and shaped after it has dried.

All of these qualities can enable a builder (or artist) to be creative in using hempcrete in all kinds of ways.

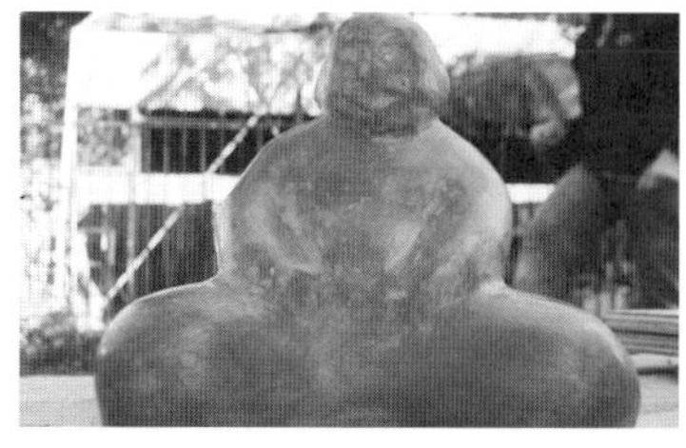

These transoms were created by molding hempcrete to support glass bottles. The material holds its shape immediately, and can support the weight of the bottles without reinforcement. Leftover hempcrete from a job site was quickly sculpted into this Buddha. The material can be a sculptor's dream!

Chapter 14

Tools

WORKING WITH HEMPCRETE insulation does not require any specialty tools; the typical tool kit of a homebuilder should suffice.

Mixing machines (both horizontal- and vertical-shaft mortar mixers) can be rented as needed. These machines keep their value on the used market, so for a larger job it can be financially feasible to buy a used machine for the duration of the work and then sell it for the same price.

It may be valuable to purchase a wood moisture equivalent (WME) probe meter to test hempcrete for dryness. Most suppliers of tools for cabinetmakers and wood workers will offer versions of this meter.

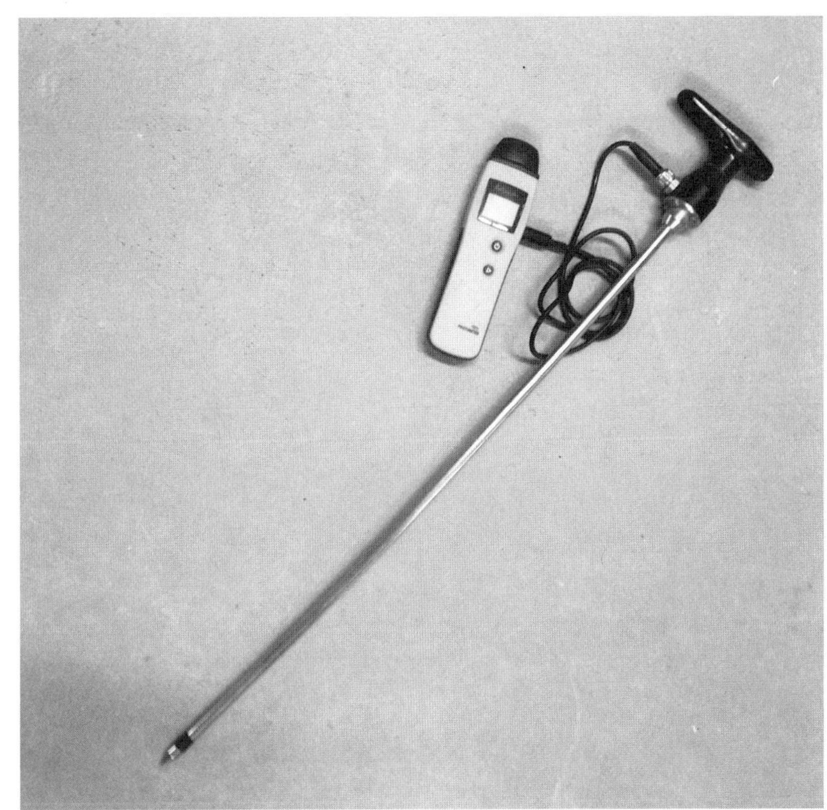

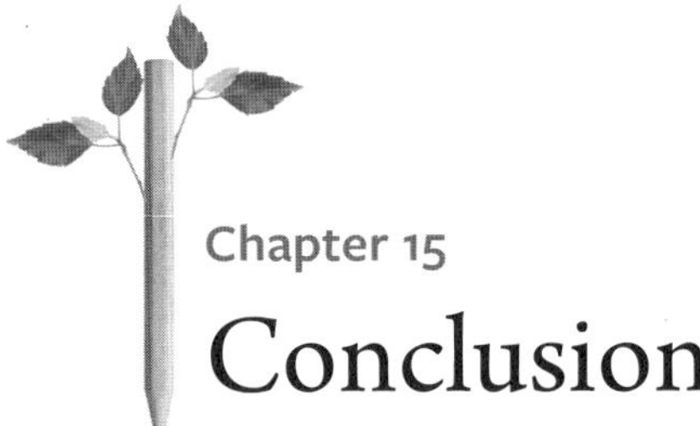

Chapter 15

Conclusion

THE CONSTRUCTION INDUSTRY is on the verge of changes from multiple factors. There is pressure to begin seriously reducing the industry's carbon footprint, to address more holistic ways of understanding and measuring the dynamic thermal performance of buildings, to create more resilient buildings that are not harmed by changing moisture and humidity conditions, and finally to create buildings that are nontoxic to occupants and the ecosystem. Hempcrete is well suited to provide a simple insulation solution for all of these changes.

The hurd used for making hempcrete is the by-product of growing industrial hemp for fiber and seed. There is plenty of evidence that this crop can play an important role in a renewed and more ecologically sound model of agriculture in North America. The fact that a great insulation material can be made from the "leftovers" of this industry is just one element of what makes hemp farming an attractive option. I am hopeful that the numerous positive aspects of hemp farming and hempcrete insulation will motivate governments, industry and citizens to push forward in this direction.

In the meantime, there is enough accessibility to North American-produced raw materials for early adopters of hempcrete to start making more living examples of these buildings. Nothing helps more to promote a new way of building than beautiful, affordable and high performing real-world examples. All the theory, promise and statistics in the world can't replace something that makes people say, *Wow.* It does all supporters of hempcrete building well when the buildings that get made are outstanding.

Hempcrete feels like the future!

Advocates of building with hempcrete need to remember that this material is not the single solution to all of the world's problems either. It is an insulation material, and it must be incorporated into well-designed and well-made buildings in order for its benefits to be meaningful. Like all building materials, it has limitations and drawbacks, and we should be open and honest about these when promoting the idea. It is easy to fall into the realm of 2G2BT — Too Good To Be True. It is a disservice to potential designers, builders and homeowners if they are not given the full picture. There are enough positive aspects to building with hempcrete that it makes an attractive choice even when its limitations are made clear.

As a builder, I am excited by the ease with which hempcrete can be incorporated into any builder's style of construction. It is rare that an ecologically friendly alternative fits so well into the existing methods of building. For this reason, I am hopeful that more builders will be

It's fun to learn how easy it is to use hempcrete.

willing to adopt this material into their "toolbox" of options.

There is something very satisfying about standing back and looking at a newly formed hempcrete wall. I would also find it very satisfying to stand back and look at hundreds or even thousands of new hempcrete homes!

Resources

A comprehensive list of books, Websites, human resources and testing.

Books

Stanwix, William and Alex Sparrow. *The Hempcrete Book: Designing and building with hemp-lime.* Green Books, Cambridge, 2014.

Bevan, Rachel and Tom Woolley. *Hemp Lime Construction: A guide to building with hemp lime composites.* HIS BRE Press, Bracknell, UK, 2008.

Allin, Steve. *Building with Hemp,* 2nd edition. SeedPress, 2012.

Amziane, Sofiane, ed. *Bio-aggregate-based Building Materials: Applications to hemp concretes.* Wiley-ISTE, 2013.

Kennedy, Joseph, ed. *The Art of Natural Building: Design, construction, resources.* New Society Publishers, 2015.

Magwood, Chris. *Making Better Buildings: A comparative guide to sustainable construction.* New Society Publishers, 2014.

North American Hempcrete Builders/Designers

The Endeavour Centre. endeavourcentre.org. Offers hempcrete workshops, design of hempcrete buildings and building of public facilities using hempcrete and othersustain able materials. Based in Ontario, Canada.

Hemp Technologies Collective. hemp-technologies.com. Workshops; material and labor supply for hempcrete homes. Based in Manitoba, Canada, and North Carolina, USA.

Alembic Studio. alembicstudio.com. Design, build and provide hempcrete information. Based in North Carolina, USA.

Hempcrete Natural Building Ltd. hempcrete.ca. Home design and construction, material supply, workshops. Based in British Columbia, Canada.

ArtCan. hempconstruction.com. Home building, consultations, technical assistance, equipment rentals. Based in Quebec, Canada.

Art du Chanvre. artduchanvre.com/english. Home design and construction. Based in Quebec, Canada.

Hempcrete Material Suppliers

BioFiber Converters
2916 – 5 Avenue NE #12
Calgary, Alberta Canada T2A 6K4
Toll Free: 1 855 984-5427
info@justbiofiber.com

Plains Hemp. plainshemp.com. Canadian-grown hemp hurd supplier.

Tradical Hemcrete. americanlimetechnology.com/tradical-hemcrete. US-based importer of proprietary hemp-lime building system.

Batichanvre Binder. limes.us/products/batichanvre. US importer of St. Astier hempcrete binder.

Rosendale Cement. rosendalecement.net. US manufacturer of natural cement binder.

Original Green Distribution. original green distribution.com. US supplier of proprietary hempcrete blend called HempStone.

Geopanel. geoproductscorp.com/gpr/prodotti/casseforme/geopanel.html. Reusable modular formwork.

Hemp Industry Resources

National Hemp Association. national hempassociation.org . US trade association for the hemp industry, with a strong interest in the use of hemp as a building material.

Canadian Hemp Trade Alliance. hemptrade.ca. Canadian trade association for the hemp industry.

Plastering Resources

Weismann, Adam and Katy Bryce. *Clay and Lime Renders, Plasters and Paints: A how-to guide to using natural finishes.* UIT Cambridge, 2015.

Guelbert, Cedar Rose and Dan Chiras. *The Natural Plaster Book: Earth, lime and gypsum plasters for natural homes.* New Society Publishers, 2002.

"The Art and Science of Natural Plaster." DVD. Chris Magwood. 2015, Montreal, Quebec, Plasterscience.com.

Frame Construction Resources

Thallon, Rob. *Graphic Guide to Frame Construction.* Taunton Press, 2002.

Canada Mortgage and Housing Corporation. *New Canadian Wood Frame House Construction,* 3rd edition. CMHC, 2006.

Simpson, Scot. *Complete Book of Framing: An illustrated guide for residential construction,* 2nd edition. RSMeans, 2011.

Editors of Fine Homebuilding. *Framing Floors, Walls and Ceilings.* Taunton Press, 2015.

Case Studies

Leskard, Marta. (2015) "A sustainable storage solution for the Science Museum Group," *Science Museum Group Journal* 4, Autumn 2015.

"Final Report on the Construction of the Hemp Houses at Haverhill, Suffolk," Client report number 209-717 Rev.1, BRE, 2002.

"Thermographic Inspection of the Masonry and Hemp Houses at Haverhill, Suffolk," Client report number 212020, BRE, 2003.

Academic Research

Lawrence, M., et al. (2013) "Hygrothermal performance of bio-based insulation materials," *Proceedings of the Institution of Civil Engineers: Construction Materials* 166 (4):257–263.

Gandolfi et al. (2013) "Complete chemical analysis of Carmagnola hemp hurds and structural features of its components," *BioResources* 8 (2):2641–2656.

Colinart, Thibaut, et al. (2013) "Experimental study on the hygrothermal behavior of a coated sprayed hemp concrete wall," *Buildings* 3, 79–99.

Pervaiz, Muhammad, et al. (2003) "Carbon storage potential in natural fiber composites," *Resources Conservation and Recycling* 39 (4):325–340.

Evrard, Arnaud and De Herde, André. "Bioclimatic envelopes made of lime and hemp concrete." CISBAT 2005 *Renewables in a Changing Climate: Innovation in Building Envelopes and Environmental Systems* (EPFL, Lausanne, Switzerland, 09/28/2005 to 09/29/2005).

Collet, F., et al. (2013) "Comparison of the hygric behaviour of three hemp concretes," *Energy and Buildings* 62, 294–303.

Dhakal, Ujwal. (2016) "The effect of different mix proportions on the hygrothermal performance of hempcrete in the Canadian context," graduate research project at Ryerson University, Toronto.

Pinkos, Jeremy J. and Dick, Kris J. (2011) "Load and moisture behaviour of Manitoba hemp for use in hempcrete wall systems in a northern prairie climate," The Canadian Society for BioEngineering, Paper No. CSBE11-24.

Kidalova, Lucia, et al. (2011) "MgO cement as suitable conventional binders replacement in hemp concrete," *Pollack Periodica* November, 2011.

Evrard, A., et al. (2006) "Dynamical interactions between heat and mass flows in lime-hemp concrete," *Research in Building Physics and Building Engineering,* Fazio, Ge, Rao and Desmarais, eds., Taylor & Francis Group, London.

Collet, Florence, et al. (2012) "Experimental investigation of moisture buffering capacity of sprayed hemp concrete," *Construction and Building Materials* 36, 58–65.

Arnaud, Laurent, et al. (2012) "Experimental study of parameters influencing mechanical properties of hemp concretes," *Construction and Building Materials* 28, 50–56.

Colinart, Thibaut, et al. (2013) "Experimental study on the hygrothermal behavior of a coated sprayed hemp concrete wall," *Buildings* 3, 79–99.

Fire Resistance Test in Accordance with ASTM E119-14, Standard Test Methods for Fire Tests of Building Construction and Materials, and CAN/ULC S101-07, Standard Methods of Fire Endurance Tests of Building Construction and Materials. (2015) QAI Laboratories Test Report #T1007-1.

Busbridge, Ruth. (2009) "Hemp-clay: An initial investigation into the thermal, structural and environmental credentials of monolithic clay and hemp walls," graduate research project at the Centre for Alternative Technology, Wales.

de Bruijn, Paulien. (2008) "Hemp Concretes: Mechanical properties using both shives and fibres," Graduate thesis, Swedish University of Agricultural Sciences, Alnarp.

Daly, Patrick, et al. (2009) "Hemp lime bio-composite as a building material in Irish construction," Environmental Protection Agency STRIVE Report 2009-E T-DS-2-S2.

Duffy, E., et al. (2013) "Hemp-lime: Highlighting room for improvement," *Civil and Environmental Research* 4, 2013.

Chew, Priscilla, et al. (2008) "Compressive strength testing of hemp masonry mixtures," graduate research project, Queen's University, Kingston.

Latif, E., et al. (2015) "Moisture buffer potential of experimental wall assemblies incorporating formulated hemp-lime," *Building and Environment,* 93 (2):199–209.

Colinart, Thibaut, et al. (2013) "Hygrothermal behaviour of a hemp concrete wall: Comparison between experimental and numerical results," *Proceedings of BS2013: 13th Conference of International Building Performance Simulation Association,* Chambéry, France.

Lawrence, M., et al. (2012) "Hygrothermal performance of an experimental hemp-lime building," *Key Engineering Materials* 517, 413–421.

Lawrence, M., et al. (2013) "Hygrothermal performance of bio-based insulation materials," *Proceedings of the Institution of Civil Engineers: Construction Materials,* 166 (4):257–263.

"Industrial hemp in the United States: Status and market potential," United States Department of Agriculture Report, 2000.

Stevulova, Nadezda, et al. (2013) "Lightweight composites containing hemp hurds," *Procedia Engineering* 65, 69–74.

de Bruijn, Paulien. (2012) "Material properties and full-scale rain exposure of lime-hemp concrete walls: Measurements and simulations," Doctoral thesis, Swedish University of Agriculture, Alnarp.

Latif, E., et al. (2015) "Moisture buffer potential of experimental wall assemblies incorporating formulated hemp-lime," *Building and Environment,* 93 (2):199–209.

Ronchetti, Paolo. (2007) "The barriers to the mainstreaming of lime-hemp: A systemic approach," Graduate thesis, Dublin Institute of Technology.

Maalouf, C., et al. (2011) "Numerical investigation of heat and mass transfer in flax and hemp concrete walls," *Proceedings of Building Simulation 2011: 12th Conference of International Building Performance Simulation Association,* Sydney.

Pinkos, Jeremy. (2014) "The effectiveness of hempcrete as an infill insulation in the prairies compared to a standard building based on power consumption," Graduate thesis, Department of Biosystems Engineering, University of Manitoba, Winnipeg.

de Bruijn, Paulien, et al. (2013) "Moisture fixation and thermal properties of lime–hemp concrete," *Construction and Building Materials,* 47, October 2013, 1235–1242.

Collet, Florence, et al. (2013) "Comparison of the hygric behaviour of three hemp concretes," *Energy and Buildings* 62, July 2013, 294–303.

Miskin, Naomi. (2010) "The carbon sequestration potential of hemp-binder: A study of embodied carbon in hemp-binder compared with dry lining solutions for insulating solid walls," Master's thesis, Centre for Alternative Technology, Wales.

Sinka M., et al. (2010) "Sustainable thermal insulation biocomposites from locally available hemp and lime." *Environment, Technology, Resources.*

Mukherjee, Agnita. (2012) "Structural benefits of hempcrete infill in timber stud walls," Master's thesis, Department of Civil Engineering, Queen's University, Kingston.

Tran Le, A.D., et al. "Transient hygrothermal behaviour of a hemp concrete building envelope," *Energy and Buildings,* 42 (10) October 2010, 1797–1806.

Index

Page numbers in *italics* indicate tables.

About the Author

Chris Magwood is obsessed with making the best, most energy efficient, beautiful and inspiring buildings without wrecking the whole darn planet in the attempt.

Chris is currently the executive director of The Endeavour Centre, a not-for-profit sustainable building school in Peterborough, Ontario. The school runs three full-time, certificate programs: Sustainable New Construction, Sustainable Renovations and Sustainable Design, and it hosts many hands-on workshops annually.

Chris has authored numerous books on sustainable building, including *Making Better Buildings* (2014), *More Straw Bale Building* (2005) and *Straw Bale Details* (2003). He is co-editor of the Sustainable Building Essentials series, and is a past editor of *The Last Straw Journal,* an international quarterly of straw bale and natural building. He has contributed articles to numerous publications on topics related to sustainable building and maintains a blog entitled "Thoughts on Building."

In 1998 he co-founded Camel's Back Construction, and over eight years helped to design and/or build more than 30 homes and commercial buildings, mostly with straw bales and often with renewable energy systems.

Chris is an active speaker and workshop instructor in Canada and internationally.

If you have enjoyed *Essential Hempcrete Construction*, you might also enjoy other

Books to Build a New Society

Our books provide positive solutions for people who want to make a difference. We specialize in:

Food & Gardening • Resilience • Sustainable Building
Climate Change • Energy • Health & Wellness • Sustainable Living

Environment & Economy • Progressive Leadership • Community
Educational & Parenting Resources

New Society Publishers

ENVIRONMENTAL BENEFITS STATEMENT

New Society Publishers has chosen to produce this book on recycled paper made with **100% post consumer waste,** processed chlorine free, and old growth free.

For every 5,000 books printed, New Society saves the following resources:[1]

27	Trees
2,407	Pounds of Solid Waste
2,649	Gallons of Water
3,455	Kilowatt Hours of Electricity
4,376	Pounds of Greenhouse Gases
19	Pounds of HAPs, VOCs, and AOX Combined
7	Cubic Yards of Landfill Space

[1]Environmental benefits are calculated based on research done by the Environmental Defense Fund and other members of the Paper Task Force who study the environmental impacts of the paper industry.

For a full list of NSP's titles, please call 1-800-567-6772 *or check out our website* at:

www.newsociety.com